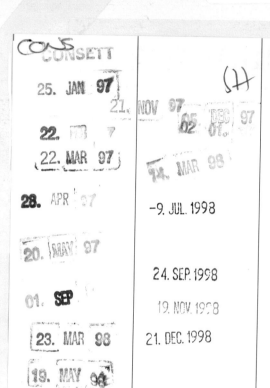

Please return/renew this item by the last date shown

DURHAM COUNTY COUNCIL
ARTS, LIBRARIES AND MUSEUMS

THE RUBBISH ON OUR PLATES

by Fabien Perucca
and Gérard Pouradier

translated by Joe Laredo

First published in Great Britain 1996
by Prion Books Ltd,
32-34 Gordon House Road,
London NW5 1LP

First published in France by
Editions Michel Lafon, 1996
as *Des Poubelles dans nos assiettes*

Copyright © Editions Michel Lafon 1996

All rights reserved.
No part of this book may be reproduced, stored in a retrieval system, or transmitted in any form or by any means, electronic, mechanical, photocopying, recording or otherwise, without the prior written permission of the publisher and copyright owners.

A catalogue record of this book can be obtained from the British Library

ISBN 1-85375-223-1

Cover artwork and design by Bob Eames

Printed and bound in Great Britain by
Biddles Ltd, Guildford & Kings Lynn

To Marie and Christiane, of course

THE MENU FOR STARTERS

A table of contents, including sweet and cheese.

Forewarning		13
1	Something of an Hors d'Oeuvre	15
2	Tea Party on a Radioactive Volcano	25
	An Active Future	30
	'Minamata Mon Amour'	33
	An Italian Job	34
	Gas Chambers on Every Floor	35
3	Steely Skies and Leaden Rivers	37
	Death by Saturnalia	38
	Dying to Get There	39
	Cadmium Cells	41
	From Aluminium to Alzheimer's	42

4	**The Setting Sun is the Colour of Farmers' Blood**	45
	Profusion Through Perfusion	46
	An Overdose of Fertilisers	47
	A Deadly Cocktail	49
	Wood for Making Coffins	51
	Grapes of Wrath	52
	Playing with Water	53
	A Packet Full of Potions	54
	DDT and Old Lace	57
	Raking it In	58
5	**Showdown at the OK Corral**	61
	Calving them Up	63
	Multiplication Stables	65
	Meating Resistance	66
	The Steak Connection	68
	Oh for the Outdoor Life!	69
	Cowboy Practices	71
	Cows that didn't Jump Over the Moon	73
	Rolling in the Isles	75
	True British Phlegm	77
	Firing Rubber Bullets	78
	Ear Piercing Cows	80
6	**The Power of the Imagination**	83
	Funny Young Birds	86
	Scheduled for Slaughter	87
	A Tocsin for a Toxin?	89
	One Man's Meat...	90
	Waste Not Want Not	91
	Grist to the Mill	92
	Soya Sauces	95
	Return to Sender	97

7	**You Can't Make an Omelette Without Breaking Eggs**	**99**
	Deus ex Machina	100
	Sure as Eggs	101
	Ovoproduction and Fishy Business	103
	Keeping us Sweet	103
	Giving it Some Stick	104
	Up a Gum Tree	105
	If at First You Don't Succeed…	107
8	**E is for Additives**	**109**
	Colours to Taste	111
	Making Oranges Orange	112
	The Acid Test for French Dressing	115
	All Roads Lead to Aroma	116
	Grave New World	119
	Glutamate and Tapioca	120
9	**Extrusion-Cooking**	**123**
	Meatless Meatballs	124
	Creating Solutions	126
	Hamming it Up	127
	Separate Ways	129
	Steep Bill of Health	131
	All Sweetness and Light	133
	Earning a Crust	134
	Indigestion	136
10	**Sowing Against the Grain**	**139**
	Science Knows Best	140
	Peaches and Crime	142
	Seek and Ye Shall Find	144
	No Fun on the Farm	145
	Mutatis Mutandis	146

11	**Alimentary, My Dear Watson**	149
	Nothing New Under the Sun	151
	More Mice than Cats	153
	The Art of Persuasion	154
	Nothing to Declare	155
	Don't Worry, it's Nothing Serious	156
	Coffins on Wheels	158
	The Garden of Eden Goes West	159
12	**Pushing the Trolley out for Sheep's Eyes**	163
	Seeing Red	164
	Organic Panic	166
	Making Mountains out of Molehills	167
	Busting a Gut	169
	Nutritional Claims	170
13	**In Vino Veritas**	173
	Vile Vines	174
	The Rot Sets In	176
	Name Your Poison	177
	Making it Go Further	177
	Scraping the Barrel	179
	Labels that Won't Wash	180
	Your Health!	182
	With or Without Gas?	183
14	**Mind the Cold!**	185
	Managing to Keep Cool	185
	Frozen Stiff	187
	Going to Work on Some Eggs	189
	Pull the Udder One!	190
	Plumbed Tomatoes	193

15	Mind the Heat!	195
	But is it Art?	196
	In the Soup	198
	'Shut Up and Eat Up!'	200
16	A Quick Dig in the Ribs to Help it All Down	203

To Finish	213
Adding Up the Bill	215
Some Recommendations	225
Don't Forget the Tip!	227

FOREWARNING

In the closed world of industrial food production there is an uneasy relationship between commercial interests and trade secrets on the one hand and the right of access to information on the other.

In France, the financial stakes are undoubtedly high. Throughout the world the label 'produce of France' is synonymous with luxury, refinement and taste, or at the very least quality. So much so that agricultural foodstuffs are France's premier industry: not only does it constitute the major share of the home economy, it is also the mainstay of external trade and one of the largest export industries in the world. So it's hardly surprising that its machinations are more or less top secret, if not a matter of national security, and that a rule of silence is strictly enforced. If for no other reason than to conceal the dreadful truths which we shall now expose.

Just occasionally a company boss, a developer or a senior administrator in some organisation we approached did welcome us with open arms, but they were the ones who had nothing to hide and we duly acknowledge them at the back of this book.

As a result, much of our information had to be obtained by extremely dubious, not to say indiscreet means. Which is why we have chosen to adopt a scorched earth policy and have declined to quote the names of most of the companies, brands and people concerned.

Our detractors may accuse us of muck-raking. We'll go along with that. After all, we've been through their bins and we can prove that what they're hiding in there doesn't exactly smell of roses...

1

SOMETHING OF AN HORS D'OEUVRE

Man may not be such a goose after all, and the worst pigsty may not be where you think it is.

One of the last jewels in the British imperial crown will soon cease to sparkle and will rejoin the former Middle Kingdom: Hong Kong – one of the world's most beautiful cities with its outrageous skyscrapers disappearing into the warm Kowloon mists.

To the north of Hong Kong and close to this increasingly temporary border, Canton is the obligatory entry point into darkest China and the mysteries of the Forbidden City. To the north of Canton stretches a vast country the size of a continent, centuries old and one of the cradles of civilisation. When the year of our Lord 2000 arrives at the door of our supposedly ancient Europe, Chinese clocks will be ringing in the Buddhist year 2543.

THE RUBBISH ON OUR PLATES

First stop on the traditional tourist itinerary is the 'hot dog' market which runs alongside the Hong Kong-Canton railway line. It has been described as the biggest attraction east of Disneyland.

The smell is all-pervasive. The persistent odour of Chinese cabbage rotting on pavements and rooftops mingles with the stench from the dog stalls where canine entrails steam gently in the chilly early morning air. One of the stalls, which extend for more than a hundred metres along the track, is advertising 'Dog Macs'.

Inside their metal cages, awaiting slaughter, man's best friends howl for their lives. Large or small, black or white, long-eared or short-eared, male or female, the same fate awaits them all: to be grabbed by a metal hook to the amazement and horror, not to say revulsion of Westerners who cover their ears so as not to hear the poor creatures' cries of pain.

That doesn't stop them taking a few photographs of course, so that back home they can ease their consciences at witnessing this Pekinese barbarity. They can then assume a pained expression and whisper, 'You won't believe it, but there's worse...'

'No!' their friends will exclaim.

'Take the swallow's nest, for example. It's not fit for cooking unless it's freshly made and the bird was so exhausted from building it that the saliva it used to stick the blades of grass together is mixed with blood. Apparently, that's what gives the nest its distinctive flavour... Then, of course, there are the fish that are still wriggling when they're served to you!'

At which point everyone will shriek in disgust at such culinary perversion.

Back in Canton, whenever a customer turns up at the dog stall, all the butcher has to do is give the animal a good bash with a hammer and slit its throat with a long knife while an old woman or perhaps an infant holds out a pink plastic bowl to catch the blood. It has to be said that dog's blood has medicinal properties which are quite unknown in the West.

When the customer (usually some modern mandarin's chef de cuisine) departs carrying the bloody carcass wrapped up in a plastic bag, ordinary housewives look on enviously: dog meat is a delicacy they will never be able to afford.

A short distance away there is a village, a traditional settlement where peasants raise prosperous, corpulent pigs – pigs whose every part, from head to trotter, is edible: you can start at either end. Isn't there an old Chinese proverb: 'There's nothing bad about a pig'? But which worships the other more, the man or the pig?

It may seem a ridiculous question. Until you visit a local peasant's house, so elegantly perched on its piles, with a great hole in the middle to allow man and beast to communicate freely. Through this hole the man pours everything he can to give the pig the best possible nourishment: dishwater, leftovers, rotten potato peelings, bones, feathers and so on.

Beneath the house, the animal wallows about

from dawn till dusk in its sweet-smelling sty. This strange bond between man and beast – a kind of pact based on the eternal principle of reincarnation – is renewed at every hour of the day. Especially after meals. And since the Chinese eat at all hours of the day, the pact is never broken and the pig never unhappy.

But it's at daybreak that the relationship is at its most intimate, when the man crouches deferentially on the bamboo floor, his trousers about his ankles, and offers the fruit of his intestines to the animal waiting below, its mouth open and its eyes closed in ecstasy behind a veil of long eyelashes.

A moment of pure bliss.

And when the time eventually comes for the man to let the animal's blood, however sad he may be at having to sacrifice a creature with which he has shared so much, he can at least console himself with the knowledge that it was well looked after during its brief spell on earth.

This is China's way of life and that of its thousands of millions of inhabitants, for whom every plateful of meat is a rare feast. Never mind that in the Year of the Pig – 1995-96 by our calendar – a few extra animals were slaughtered by way of celebration.

* * *

Five thousand kilometres west of the Great Wall, in an area seventeen and a half times smaller and nineteen times less populated than China, the

SOMETHING OF AN HORS D'OEUVRE

love affair between man and beast has taken quite a different turn.

In this tiny country at the edge of the earth called France, a feast is no longer a red-letter day. For several decades now the French have made sure that they never go without meat.

To begin with, they no longer raise their own pigs, having delegated this task to experts. Secondly, they no longer kill them at home. They're not allowed to for reasons of hygiene, as is perfectly understandable. And finally, if they won't bend a knee to pay their respects to the sacrificial animal, it's because they have more important things to do. Besides, the little white tray wrapped in cling film which they toss into their supermarket trolley – in between the washing powder refill and the economy pack of natural yoghurt – offsets the flesh colour so attractively that it serves as a perfectly adequate coffin.

But the biggest difference between the Chinaman and the Frenchman is in the number of animals they kill: while the former consumes no more than twenty-five kilos or so of meat every year, the latter gobbles his way through four times as much (the UK average is 70 kilos per head each year).

So are the Chinese worse off? Not necessarily, at least when you realise that the French, even though they have survived nouvelle cuisine and are now converted to health food, no longer eat themselves to death but instead kill themselves simply by eating.

Whether this is due to pesticides, fertilisers,

additives, exhaust fumes, household waste, industrial pollution or radioactive fallout, all the studies show that the food chain has been seriously disrupted. The result is a nightmare.

People in high places are suggesting that 80 per cent of cancers are caused by what we eat.

Modern medicine is making constant advances and working all kinds of miracles, and yet there is more and more illness, albeit unseen illness. Operating theatres and chemotherapy units are full and the statistics are alarming. Every day, four hundred people die of cancer in France alone (the UK figure is 450 deaths per day). A quarter of the population either has had cancer or will have had it by the year 2000. And according to the worst predictions, the proportion will increase to a third after the year 2000 and to half by 2010.

Yet hardly a day passes without our food being inspected, checked, analysed and quality assured by the Ministries of Health, Agriculture, Trade & Industry. Not to mention the National Food Hygiene Commission and all the other private and public organisations up to and including the European Commission in Brussels and its numerous offshoots. But all their efforts are in vain.

The scale of the catastrophe is beyond them. All they can do is to establish and attempt to enforce an ADI, an 'acceptable daily intake' of chemical substances which we unwittingly consume, measured in milligrams per kilo of body-weight.

This is the state we're in. The food chain is so contaminated and commerce so blinded by the profit motive that human life is valued only in inverse proportion to the turnover of the food industry. If we are not all to poison ourselves mouthful by mouthful and end up as cancer research subjects in state institutions, we need to be able to calculate for ourselves the acceptable daily intake of toxic substances which we ingest.

Since this is practically impossible, a virtual conspiracy of silence has come to surround the ADI. Those who suspect the truth aren't allowed to find out and those who know keep quiet.

* * *

The ADI is nothing new. It was Theophrastus Bombastus von Hohenheim (aka Paracelsus), a sixteenth century alchemist and physician, who first propounded the theory that 'a poison is defined by the dosage' – the starting point for today's scientific calculations which are based on the LD, the 'lethal dosage', as tested on guinea-pigs and rats.

When they test a poison, they describe it as LD20 or LD50 or LD100. By which they mean that a certain dosage manages within a specified period of time (usually a short one) to kill 20 or 50 or a 100 per cent of the rats experimented on.

From this they deduce that an LD50 is less than 5 milligrams in the case of the most toxic substances, between 5 and 50mg for very toxic substances, between 50 and 100mg for moderately toxic substances and between 500mg and 5g for slightly toxic substances.

All that for less than a kilo of rat meat.

Then they divide by a hundred to arrive at the acceptable daily intake, which is calculated for the 'average' man – he being about forty years old, weighing 70kg and having a varied diet, a reasonably healthy lifestyle and some degree of mental stability. Everyone else – women, fat people, thin people, young people and old people –is eliminated literally, because the ADI doesn't apply to them.

And it changes from day to day according to manufacturing requirements, incidences of pollution, unnatural disasters like Chernobyl, laws and regulations discreetly proposed (or should we say strongly recommended) by the major agricultural and food manufacturing groups.

It's something which is discussed daily in letters and telexes and faxes at the lowest level (farmers and breeders) and at the highest level (processors and distributors) by all those who supply the consumer outlets, from cornershops to superstores. The ADI is even the subject of one of the food industry's many trade journals. Every debate, every convention, every seminar revolves around it.

More and more complicated chemical concoctions and increasingly frightening genetic manipulations mean that production is constantly increasing and exceeding by an ever widening margin the absolute limits. Whether it's crops or livestock, milk or wine, honey or herbs, ready meals or canteen food, potatoes or papayas, the

SOMETHING OF AN HORS D'OEUVRE

boundaries of common sense are being left further and further behind under pressure from farmers, processors, manufacturers, packagers, distributors, in fact everyone who is involved in this mad race for profit, even though they are all running towards inevitable and imminent disaster.

* * *

It's a 'gold' rush for the riches of the land. Riches worth sufficient billions to make the Chancellor beam with pride. Riches with which to terrorise governments through electoral distortion (constituency boundaries are such that a single farmer's vote in the Creuse valley carries as much weight as a thousand votes in Versailles). Riches which have made France the world's biggest producer and biggest exporter of 'processed' foodstuffs. Riches in whose name 58 million people are being poisoned, slowly but surely and without even realising it.

And that's not including all the millions in other countries who buy French produce.

It is certainly no accident that French scientists put their names to almost none of the studies into the relationship between food and health. The documents that are published all come from Belgium, Denmark, Holland, Germany, Great Britain or Sweden. In any case, it is not in the French government's interest to provide its citizens with a user's guide to the ADI – an extremely sophisticated yet at the same time entirely theoretical and therefore totally inaccurate calculating tool. Why? Because manufacturers all over Europe regularly lie about the quality of their products. Because no label can ever really tell you what it is you're eating. And because, ADI or no ADI, the situation is so desperate that no one really knows what to do about it.

So bad, in fact, that in the offices of the 'Department of Food' in central Paris, the following message is permanently displayed on every computer screen:

'Any problem can be resolved by leaving it unresolved for long enough.' (Sic!)

2

TEA PARTY ON A RADIOACTIVE VOLCANO

*Today's specials: 'Curie' rice,
Vol-au-vent in sievert sauce
and Polynuclear soup
flavoured with toxic fumes*

It is up to us to work out our own acceptable daily intakes, unless we want to die of ignorance. It's simply a matter of opening our eyes and looking around us. For a start, we all remember Chernobyl. On 26th April 1986 Reactor No.4 at the Chernobyl Nuclear Power Station in the Ukraine goes haywire. Its core collapses in a gigantic explosion fuelled by combustible radioactive material. Within a few minutes it has released into the atmosphere a mega-radioactive cloud equivalent to all the above-ground atomic explosions since Hiroshima put together. The famous cloud which we were assured had only just reached the Swiss border.

Everyone who works in the nuclear industry understands only too well what an 'acceptable intake' is. To them it's the maximum amount of radiation the body can safely absorb in a year. Previously it was calculated in curies, rads and rems; nowadays they talk in becquerels (Bq), grays (Gy) and sieverts (Sv). The average natural background radiation in France is 2 millisieverts per annum, from which it's reckoned that the average person can tolerate 5 millisieverts per annum and that an employee of the French Atomic Energy Commission can withstand ten times that amount on condition that he undergoes constant medical observation.

A dose of a tenth of a sievert (100 rems) or more has noticeable effects on the body. Three-tenths of a sievert and the blood starts to deteriorate: among other things the number of white corpuscles (the ones which protect the body against hostile invasion) diminishes rapidly.

But on that Saturday morning, 26th April, while the invisible enemy was at large, no one in France had been warned of the danger. The only news that was available came from foreign radio broadcasts. In Germany, Switzerland, Sweden, Denmark and Norway people were barricading their homes, eating out of tins and drinking only bottled water. Hospitals received thousands of calls. More than a hundred thousand women across Europe terminated their pregnancies as a result of medical advice. The panic soon spread to Spain, Portugal and Italy. Two and a half million

Ukrainians had already been exposed to radiation. The whole of Europe was in a state of hysteria, even the Americans were worried, but in France nothing happened. The authorities kept quiet. In Paris, Lyons, Marseilles and Toulouse people went about their business as if nothing had happened.

And it wasn't until 2nd May, six days after the disaster, that the French government finally decided to inform its citizens. By then it was far too late to think of self-protection.

It was almost another month before the first official reports on the contamination of flora and fauna were published.

They revealed nothing dramatic, in fact far from it. The cloud had largely dispersed and passed harmlessly by. True, it did cross part of France, particularly the east and south-east as well as Corsica, and the atmosphere was contaminated forty times as much as in Spain but only a quarter as much as in Italy. On the other hand, the reports didn't mention exactly what had happened during the few days between 26th April and 2nd May.

In fact, on 26th April itself, the Head of Fruit and Vegetables at the 'Department of Food', Jean-Claude Evrard, was at his country house near Avignon. He was in his shirtsleeves, having a relaxed breakfast in the garden and enjoying the warm spring sunshine, when a hysterical phone call from Paris suddenly took away his appetite.

'Jean-Claude?'
'Yes, what is it?'

'Listen, there's been a disaster, we need everyone...' The voice at the other end of the line was panicky. 'Pull up some grass and some herbs and grab a handful of soil and get over to the Route des Chappes.'

Route des Chappes is the address of a laboratory in Sophia-Antipolis, two hundred kilometres from Avignon.

'That day', Evrard recalled, 'I broke all the speed limits and I'm not sorry. The results were terrifying.'

That's as much as we know.

But there are some figures available. On 26th August, exactly four months after the Chernobyl explosion, tests carried out on milk reveal radioactivity of 28,000 becquerels per litre, whereas the official limit (which was hastily increased) was 370 becquerels per litre. The next day, it is discovered that all the aromatic and medicinal herbs in the Drôme region would be contaminated for at least a year. On 3rd September, lake fishing is banned in Switzerland. From 29th September, 500,000 radioactive sheep are slaughtered in Great Britain. On 2nd October, the Malaysians send back 45 tonnes of contaminated butter which had been shipped from Holland. In 1992 a further 600,000 sheep are killed in Scotland, but in France still nothing happens.

Ten years on, trying to obtain copies of any of the official documents relating to the many studies undertaken since that time (approximately twelve per year) is like going on an assault course. You can't help wondering whether these studies are classified as secret when you hear a PR Manager for the Ministry of Agriculture exclaim, 'If we let these documents out, we will be shot.'

However, from a strictly unofficial source, we know that virtually all traces of contamination in milk, meat, vegetables and fruit had disappeared by 1989. All that was left behind were abnormally high levels of cesium 134 in mushrooms from Alsace and ferns from the Vosges.

Nevertheless, we shouldn't celebrate too soon. Firstly because, according to some scientists, those most exposed to radioactivity (i.e. those who unsuspectingly went outside their houses at any time during the six days between 26th April and 2nd May 1986), if they are going to develop cancer as a result, will not do so until after the year 2000.

Secondly because when this kind of disaster strikes, it would seem that the French authorities are not to be relied on. Nor is the European Commission, which wasted no time in moving the goalposts as far as permissible levels of contamination in food products were concerned. It was either that or close every frontier and cause commercial ruin throughout Europe.

But also because the radiation from Chernobyl is still spreading, notably via migratory birds which the French have no hesitation in shooting down.

And finally, perhaps most importantly, because there's no guarantee that there won't be another explosion at a nuclear power station. Anne Lauvergeon, who worked in the Elysée Palace for fourteen years alongside François Mitterrand and is extremely well-informed, is unequivocal: 'There is a very strong chance that another nuclear power station will blow up before the end of the century.'

A Radioactive Future

Another nuclear power station on the point of blowing up? What about Paks in Hungary or Bohunice in Slovakia or Rovno in the Ukraine or Ignalina in Lithuania? Or maybe one in central Russia or Bulgaria? There's a particularly dilapidated one in Kozloduy, for example, whose main reactor was started up again in October 1995 to the horror of scientists and governments in the West. No one knows whether the reactor vessel will hold together. If it doesn't, Western Europe will be just as seriously affected as the area around Chernobyl was in 1986.

And even if none of these stations blows up (the worst doesn't always happen), what about the thirty-one other 'Chernobyls' which have been lost at sea?

> 10th April 1963: the American nuclear submarine SSN-593 sinks just east of Cape Cod. Radiation is still escaping from it today.

25th February 1966: an American cruise missile lands in the Beaufort Sea. It is still there.

21st January 1968: a B-52 'mislays' four nuclear warheads on the ice cap. They haven't been found yet.

May 1968: another US Navy submarine, the SSN-589, sinks 400 miles off the Azores: the radioactivity of its reactor is roughly 1,295,000 gigabecquerels and it's only a matter of time before it implodes.

7th April 1989: the Soviet submarine Komsomolets hits the bottom off the Norwegian coast. Swedish cameras clearly show the rust eating away at the nuclear missiles.

14th September 1989: an American F-14 fires off a missile into the sea off Scotland.

And that's not all.

Four bombs have been lost just 10 kilometres off the Spanish coast (only one has been recovered), a Soviet submarine full of torpedoes is still lying at the bottom of the Bay of Naples, a fully armed B-47 has disappeared somewhere in the Mediterranean, an F-102 which overshot the deck of its carrier has buried itself in Haiphong Bay, and two nuclear satellites, a Russian one and an American one, have also fallen into the sea, one off Brazil, the other somewhere near Madagascar.

There are also Russian nuclear submarine bases in the Arctic where the slightest electrical fault can cause their reactors to overheat. Not to mention the thirteen nuclear reactors dumped in the Kara Sea, or the tens of thousands of radioactive waste containers lying at the bottom of oceans just about everywhere, having been disposed of by just about every country with nuclear power, which are the subject of the greatest secrecy of all. There are even some a few leagues off the French coast.

We've reached a stage where some scientists (admittedly perhaps the more eccentric ones) wouldn't altogether discount the possibility of a global atomic explosion. An overheating Russian nuclear reactor, even one at the bottom of the icy waters of the North Pole, could set off a chain reaction with an American missile lost off the Norwegian coast which in turn would prompt radioactive containers in the Golfe de Lion to disintegrate, and so on.

Fortunately, this hypothesis is unlikely to become reality. Meanwhile, polluted water flows from ocean to ocean and from hemisphere to hemisphere. And no one can be certain that radioactivity isn't building up inside the fish, just as mercury pollution at Minamata[1] spread to the whole bay and poisoned 20,000 people.

---

[1] A large port on the west coast of Kyushu island in southern Japan.

'Minamata Mon Amour'

By the time the Minamata pollution came to light in 1969, it had already caused widespread devastation. It is estimated that 30% of children born between 1956 and 1967 on the shores of Minamata Bay were affected by toxic waste from the Chisso factory. Almost a thousand people have died in terrible pain, and several thousand more have undergone dreadful suffering. Their only crime was to have swum in the sea or to have eaten fish contaminated by mercury.

Photographs of deformed children and women with limbs eaten away by disease appeared all over the world, but it didn't do any good. The red metal continues inexorably to overflow into rivers and oceans. During the past five years 2,000 tonnes of mercury have been spilled into the waters of the Amazon by psychopathic gold-diggers. Mercury can even be found in the Great Lakes of America and Canada, because as well as being carried by the sea's currents, it evaporates into the clouds and escapes into the atmosphere before falling back to earth again.

In France, where the maximum permissible amount of mercury in fish has been fixed at 0.5mg per pound, hasty samples taken in salerooms by independent analysts have revealed mercury content of up to 0.8mg per pound.

What's more, the official ADI of mercury is less than 0.05mg per kilo. After all, if it can dissolve gold and silver in a split-second, it must be able to do serious damage to the nervous system. The living dead of Minamata are evidence of that.

There's no point in panicking about it, though, because by 2010 there won't be any more contaminated fish in the sea. In fact there won't be any marine life left at all. We long since exceeded the critical annual fishing limit of 80 million tonnes, above which there are no longer enough fish left to replenish themselves. We shall soon be faced with a grave dilemma: what to give our pigs and chickens to eat when they're currently force-fed on fish meal in order to turn them into meals for the fish. Should we breed fish to feed the pigs or vice versa?

An Italian Job

While awaiting their own extinction (as a result of the greenhouse effect which threatens to put an end to evaporation by 2010), the clouds that transport the mercury also carry many other industrial pollutants. Thus in 1976 in a suburb of Milan called Seveso, several tonnes of tetrachlorodibenzo-p-dioxin (or TCDD) were released into the atmosphere within just a few minutes. It led to a mass of precautionary abortions, cost the factory owners £45 million in compensation, and forced them to decontaminate 'in depth' several hundred hectares of land.

That meant transporting more than 250,000 cubic metres of soil to a secure site, slaughtering tens of thousands of animals and putting a similar number of Italians under medical observation. What they discovered was a four-fold increase in cancer cases. And that isn't the end of it by a long way, since the dioxin they left behind will remain toxic until around 2040.

Terrified by the amount of compensation to be paid out and the premiums to be reimbursed, insurance companies went to the European Commission with a 'Seveso Directive', a particularly austere piece of writing which required manufacturers to take drastic precautionary measures. None of which has stopped factories all over Europe producing Seveso-style dioxin, nor 367 of them in France alone profiting from the aforementioned directive.

Gas Chambers on Every Floor

On the other hand, all PCBs (polychlorinated biphenyls to you and me), the most neurotoxic and carcinogenic of substances, have been banned from use in Europe. And yet you can still find them everywhere, inside those huge electricity transformers where they act as insulators, in partitioning, in the chassis of cars more than fifteen years old and in office block roofs. Eight million tonnes of PCBs have polluted Europe in the past twenty years, and they will continue to poison our planet until 2050.

And probably for longer than that since the USA, which was the first to ban PCBs at home (that was in 1978), has become the world's biggest PCB exporter. They make 100,000 tonnes of it every year and remorselessly ship it out to Africa, Asia and South America. Less familiar but just as noxious are PAHs (polynuclear aromatic hydrocarbons) which also contaminate the food chain. The only PAH to have achieved a certain notoriety is benzopyrene, the principal carcinogenic in cigarette smoke. But the majority of it (that's to say 80 per cent of the benzopyrene each of us absorbs every day) comes from our food. PAHs can cause cancer by means of inhalation, ingestion or just skin contact. Their main targets are the lungs, the stomach, breasts, bones and skin. The principal cause of benzopyrene contamination is exhaust fumes which settle both on land and in water.

And now a short domestic science lesson for environment-conscious non-smokers: even those addicted to the vice are a hundred times more likely to get cancer from exhaust fumes than from smoking. It's a figure worth bearing in mind next time you get into your car or board an aeroplane. And an all no-smoking Jumbo jet causes more pollution on take-off than a thousand cars jammed up on the motorway.

This sort of pollution of the food chain, especially by car drivers, is all the more inexcusable when you think that everything we've talked about, from Chernobyl to polynuclear aromatic hydrocarbons, has been widely reported in the media.

3

STEELY SKIES AND LEADEN RIVERS

You won't get an iron constitution from nut salads and stir-fried pig with smelted butter.

A more surreptitious though no less deadly form of pollution is caused by heavy metals. These have one amusing characteristic in common: they can't be broken down inside the body. Instead they club together and form a concentrated union in order to do us even more harm. Our Lord and Creator, if He exists, certainly couldn't have imagined how these metals could end up on our plates. Only our manufacturers know: car and street lamp manufacturers, pipe and battery manufacturers, frying pan and, of course, food packaging manufacturers. But whether through stupidity, thoughtlessness or cynicism, they do nothing to limit the production or the use of these metals.

You'd think we had a collective death wish which was driving us all towards the edge of the precipice.

Death by Saturnalia

The maximum permitted dose of lead in food is 3mg per week, but it has a history of 'gastronomic' use which is as old as civilisation. Roman viticulturists used it to 'sweeten' their wine. It is even said that every time an amphora was placed on the patrician's marble table, he would swallow as much as 800mg of lead per litre. As a result, there are those who venture the claim that it was lead poisoning which brought about the decline of an Empire that had lasted for more than eight hundred years.

If the story is true, it goes to show just what a toxic substance lead is. But it was this very property which gave rise to its reputation. The Romans called it 'saturn' after the planet they believed to be made of lead, and the Saturnalia became synonymous with the famous Bacchanalia – an orgy of debauchery and disorder in honour of the 'erratic' planet. And the disorder they worshipped is the very same disorder the body suffers as a result of absorbing lead: blood poisoning, anaemia, colic (the notorious 'lead colic') and motor damage. Tremors and 'claw hands' can also be caused by lead.

It is therefore no surprise that the first occupational disease to be recognised was saturnism in painters, who spent all day mixing lead-based paints. Decades later, their paint continues to cause damage in old houses where the passing years have reduced it to dust, if it hasn't already flaked off in minute particles which have polluted the entire kitchen, from saucepans and plates to glasses and cutlery. And we haven't even mentioned the lead piping hidden away inside walls which daily concentrates the poison in our drinking water.

You can't see, taste or smell lead particles, but you can die from them. In 1994, within the city limits of Paris, 3,000 children became sick from lead poisoning

Dying to Get There

To add to yesterday's pollution, there's today's plague – the flood of cars. In Paris alone, car exhausts emit a tonne of lead into the atmosphere every day. The organic derivatives of lead which are incorporated into petrol are a hundred times more dangerous than the metal itself and can directly attack the nervous system. In big cities, traffic jams have become just another method of suicide.

The weaker ones, who aren't so good at digesting their daily dose of black smoke, succumb early. Research undertaken in Belgium and Denmark has revealed traces of lead in babies' hair and teeth. The stronger the concentration of lead, the weaker the intellect of the child. In France it has even been found that the IQ of children born to mothers who have been subjected to high concentrations of lead is significantly lower than that of other children.

Another study carried out in Taiwan has shown that children exposed to lead pollution develop an alarming tendency towards anarchic hyperactivity, a common disorder in the USA where it manifests itself in inattention, irritability and often aggressive behaviour. In every case there is worrying evidence of retarded growth.

The faster you drive, the more petrol you burn and the more lead is emitted. What's more, being a heavy metal, it doesn't vaporise and kindly disappear into the air: it simply falls back down beside the road or the motorway, as often as not in a field. As they watch the cars go by, those pretty cows are being poisoned without even realising it: there's lead in their livers as well as their milk.

There is even more in certain types of crop which readily absorb metals, like basil, mint and rosemary, spinach, carrots, cabbage, potatoes and radishes. And even in honey, if it comes from hives which are near large towns.

So-called 'unleaded' petrol and catalytic converters aren't enough to reverse the trend, since 'green' fuel only cuts pollution by 40 per cent. Only the introduction of a universal 40 mph speed limit would have any significant effect.

On top of all this there are the great 'protectors of the environment', hunters, who scatter lead pellets in places even the four-wheel drives can't reach. Every season they fire off hundreds of kilos of the stuff, which lands on the ground and penetrates the soil when it rains or else litters the lake and river beds before being swallowed by geese and ducks, which then die in less than three weeks. Except in countries where lead pellets are illegal, i.e. Australia, Canada, the USA, Denmark and Holland. Conspicuous by its absence from this list is France, which gets through 250,000 tonnes of lead every year.

A further 120,000 tonnes is contributed annually by discarded car batteries – all lead-based. How can such a plague be checked?

Cadmium Cells

Far less familiar than lead because it has only been in use for a short time, cadmium has now joined the ranks of the great food pollutants. Lionelle Nugon-Baudon, of the 'National Institute of Agronomic Research' (INRA)[1], puts it in a nutshell: "There is contamination everywhere and it will get worse and worse unless something is done soon. But if I had to choose a single substance which required twenty-four hour surveillance, it would be cadmium. That isn't paranoia: all the experts agree."

This soft, bluish-white metal which is used for painting cars, stabilising certain plastics and manufacturing cells and batteries, is actually far more toxic even than lead. It passes from soil to

---

[1] And author of an excellent book called *Toxic Bouffe* (Lattès-Marabout, 1994).

crop with alarming ease and is particularly attracted to the liver and the kidneys. It isn't yet to blame for all the dialysis machines which hospitals are buying, but it can already be held responsible for a considerable number of cirrhoses, even in patients who have never touched a drop of alcohol.

Whereas the use of lead is in decline, the use of cadmium is constantly increasing. At the beginning of the century, less than 20 tonnes of it were produced in a year. More than a thousand times that amount will be produced in the year 2000. By then the damage will be irreparable, especially in France, the world's fourth largest consumer.

No one is safe. Cadmium diffuses into both air and water and lurks in the offal of animals, in mussels and oysters as well as in the veins of plant leaves (including salad vegetables) and finds its way onto our plates without having lost so much as an iota of its noxious strength.

In France, just the tiny button-batteries used in watches and electrical equipment release 90 tonnes of cadmium into the environment every year – unless they are recycled.

But they aren't. Or so few of them that it isn't even worth talking about.

From Aluminium to Alzheimer's

That isn't the worst of it. Take aluminium, for example, which since 1972 has been known to be one of the principal causes of premature senile dementia, or Alzheimer's disease.

There were suspicions even before that, when it was observed that communities living in areas where the soil is rich in bauxite (the mineral from which aluminium is extracted) were more prone to the illness than others. It was even noticed that in villages where the well water had more than 2mg of bauxite per litre, the number of old people 'going gaga' was well above the national average. From which these early investigators concluded that aluminium was one of the most dangerous metals and a serious health hazard. Indeed, in the human body, cancerous cells contain a far higher proportion of aluminium than healthy ones.

And every year more than 100 million tonnes of aluminium is produced. What for? For cars, cars and more cars, but also for saucepans, tins and girders, and in different alloys with so many uses that there are even traces of it to be found in baby food.

But the ultimate insanity of the aluminium craze is the 'tin' can, the celebrated drink can which has overrun supermarkets everywhere. Europe alone makes 24 billion of them every year.

It was the Swedes, one of the big can producers, who invented the layer of lacquer which separates the metal container from the liquid contents. The official reason for it was so that the can didn't make the drink taste tinny; unofficially, they acknowledge that it was mainly because of the health risk. American research has so conclusively proved it that from now on even tins of pet food have to have lacquer inside.

One thing we can hope for is a lull in aluminium production as we approach the end of the century. Not for health reasons – let's be realistic! – but due paradoxically to the overproduction of steel in industrialised countries, including Taiwan and South Korea. The cost of steel is falling while that of aluminium is rising, so that aluminium products will become less profitable in future.

But the pendulum will undoubtedly swing back the other way. While steel makers are now marketing a 32cl can weighing 18g, the aluminium makers are aiming at one weighing just 12g. In this sort of technological battle the deciding factor will inevitably be economic rather than environmental. Alzheimer's disease carries little weight against multi-million dollar profits.

The only consoling thought is that, compared with the big bad three – lead, cadmium and aluminium – the other metallic substances which find their way onto our plates are almost benign. Tin behaves like lead, high doses of fluorine cause osteosclerosis (abnormal hardening of bone or bone marrow), germanium attacks the nervous system, vanadium (a by-product of fossil fuel combustion) is another contributor to Alzheimer's disease, zinc causes malformations in embryos, platinum is carcinogenic and selenium is highly poisonous.

Mere everyday trifles.

4

THE SETTING SUN IS THE COLOUR OF FARMERS' BLOOD

Wherein we discover that agriculture is the cash cow of the chemical industry, which milks it for all it's worth.

If you were dropped in a farmer's back yard, you might have to pinch yourself to believe what you saw. That is if you were even allowed to look, because there's nothing more hush-hush than the strange chemistry which goes into growing peas and beans these days. Farmers aren't so stupid and they certainly don't want the public to know how 'nature's miracles' keep happening year after year.

Yet the evidence is all there, scattered carelessly among the massive agricultural machinery costing hundreds of thousands, the spare tyres and the galvanised tanks, there inside the breeze-block and corrugated iron sheds.

To start with, there are the 50 kilo seed bags marked, in huge letters: 'Not for human consumption. Ensure seeds are not eaten by animals or birds. Do not leave on the ground. Destroy bag after use'. Or even: 'Destruction of this bag is strictly forbidden. Bury not less than 50m from housing or drinking-water supplies'.

Even the three hens scratching about between the barn and the cowshed aren't interested in this special seed. The old geese turn up their beaks in distaste, and as for the ducks – the most cautious of fowl – they won't go anywhere near it. But it is this very seed, treated with chemicals to make it 'crow-proof' (a thousand times more effective than electronic scarecrows), which produces our daily bread.

Just nearby, we find ourselves in the world of science-fiction when we come across the 'soil treatment' products. Products, in the plural, because there are many different types which are used at different times of year. They include insecticides, nematocides (for killing worms), molluscicides (to deal with slugs and snails), crow killers and repellents, mole killers, rodenticides, antiviral compounds, etc.

Sowing is no longer a natural process. It is a high wire act requiring serious study of applied chemistry.

Profusion Through Perfusion

It starts in the autumn when tankers deliver the nitrogenous fertiliser base. It has to come in tankers because farmers pour 500 kilos of it onto every hectare. Next to arrive are assorted weed-killers, usually combined into one or two 'specials' known as 'total treatments' – in other words they 'treat' everything in sight. Make no mistake, you're far more likely to see cornflowers and poppies growing round the edges of meadows than around arable land. The poor cows wouldn't last long on a diet of poppy-killer.

Then it's the turn of the fertilisers themselves, namely about 100kg of potassium and another hundred of phosphoric acid for each hectare.

Leave until the spring and then give a second dose of 'total treatment' since the first one didn't do much good. Weeds aren't stupid – it didn't take them long to invent the biochemical equivalent of the gas mask. Every year the big chemical manufacturers launch more and more powerful products, and every year, with an agonised cry of protest the earth pushes skyward tiny shoots like kamikaze commandos on a suicide mission to replenish the oxygen upon which our survival depends.

An Overdose of Fertilisers

No one would put much money on planet Earth's chances of survival, but that shouldn't stop us admiring her courage. After the weedkillers, she has to face the fungicides, another wonder of science. The job of fungicides, which are part of the great pesticide family, is to destroy the microscopic fungi which attack cereals, fruits and vegetables and to prevent their natural decay. Fungicides penetrate the young shoots and then dissolve into the soil.

The first spraying takes place in mid-March, there are two more doses in April and May and a fourth one a few days before the harvest.

That's the theory.

In practice, farmers double and even triple the dosage to increase its effectiveness without worrying about what the Law has to say. And it does have something to say – the famous acceptable daily intake – in view of the extreme toxicity of these chemicals.

You only have to read the instructions which adorn the plastic containers to be convinced of that. One of them carries the warning: 'Any unspecified usage could lead to prosecution'. Another is labelled simply: 'Ultra-poisonous' and is to be handled only by people wearing space suits (nothing less than a face mask and protective gloves will do). But you are allowed to burn the container when you've finished.

There's worse.

Parathion, for example – the 'Mr T.' of insecticides, where T is for toxic. Any more than a thousandth of a milligram per kilo of consumer (that's one part in a billion) and parathion is lethal. Fifteen Iranian children died when they were shampooed with parathion to get rid of their lice. In 1969 bread made from parathion-contaminated flour caused fifteen deaths in Mexico and sixty-three in Colombia.

So there you are – the ADI of parathion is a millionth of a gram. One litre per hectare is enough to kill anything in the vicinity that breathes. This stuff is so strong that the law says it has to be kept 'under lock and key, well away from animal feed, drink or water'. Even the plastic containers are dangerous, being so highly impregnated with it. The label makes it quite clear: 'Not for re-use. Empty completely, rinse and destroy according to prescribed regulations'.

But ask a farmer what these regulations are and he'll probably say, 'How should I know?' Not by asking the manufacturers, who incidentally do not take their bottles back.

So come the autumn, after the harvest and before the new seeds are sown, the fields are lit up by bonfires, as conscientious but disenchanted men of the earth set light to dozens of plastic containers which were cluttering up their barns. And since it doesn't matter where they do it, why not a few yards from a drinking-water supply? We've seen it happen. But then, even if it had been rinsed out three times, one of these containers accidentally used to water a vegetable patch would have something like the effect of a swarm of locusts.

A Deadly Cocktail

Have our farmers gone mad? No. They simply have no option, because this is the only way they can increase the yield and make their enterprises profitable.

A cereal farmer, for example, will use enough poison in a single season to contaminate a medium-sized town. A fruit farmer will spray his apples with five 'units' of liquid nitrogen, once in the autumn and once in the spring, plus 90 units of phosphoric acid and 110 units of potassium. He is also allowed to use four different weedkillers, since it is a well known fact that grass stunts trees. As for special treatments, twenty years ago he would have sprayed his trees with paraffin oil; nowadays he can use twenty-three different copper based fungicides, eight sulphur, six captan, five thiophanate-methyl, one thiram and one oxine-copper.

When it comes to insecticides he can choose from fourteen different treatments of phosalone, five of chinomethionat and one of tetrasul, apart from parathion. Not forgetting the specialised chemicals for 'enhancing' the colour of the fruit, like naphthaleneacetic acid or urea. That's when he isn't sprinkling his cherries with ethephon, a hormone naturally present in trees but reproduced chemically to make the fruit ripen on the exact day at the exact time the grower decides to pick it in order to get the best price from the wholesaler.

All this inevitably ends up on our plates. An innocent-looking field of strawberries disinfected with methyl bromide is a time-bomb waiting to go off, while ant solution also wipes out the rat population and causes cancer in humans. Just one milligram is enough to spark off a massive metastasis in weaker specimens; yet strawberries can contain as much as 5mg per kilo. Doses up to ten times higher have been found in greenhouse tomatoes, and it takes a month for apricots to break down even a tenth of these carcinogens.

The walls surrounding the ADI are crumbling!

Wood for Making Coffins

As with any drug, the higher the dosage, the lower the effectiveness. Spiders and aphids simply mutate and within three days are back, vaccinated, immunised and extremely hungry. The only way of getting rid of them is to start all over again, possibly changing the chemicals or more often simply doubling the amount of poison.

Let's look at one of the more unexpected examples of the use of chemicals in the farmer's day to day life: those little wooden boxes used to stack plums or apples. Even before they have left the sawmill, they're treated with pentachlorophenol (PCP), a highly carcinogenic fungicide which has already been banned from use on any interior woodwork or in anything to do with food.

It is so dangerous that within a few years its ADI has been reduced from 3 micrograms per kilo to a twelfth of that amount, yet fruit imported from Spain, Switzerland and Germany has been found to contain anything between 4 and 16 micrograms. Figures for French fruit are as well publicised as military secrets, but we can reasonably suppose that they are in line with this honest European average.

'It's not our fault', say the sawmill owners, 'We only supply the wood. We don't know what the customer uses it for!' Hand on heart, the manufacturers of the boxes protest their innocence. 'Often we don't even know whether it has been treated with PCP.'

Meanwhile, the spread of pentachlorophenol continues. Being highly volatile, it is also extremely pervasive and remains toxic for up to twenty years. Most French apples are contaminated by it, as well as baby food and other things besides.

Grapes of Wrath

Beware! Even ordinary insecticides can have unexpected side-effects. In certain parts of France wine growers have stopped using vine shoots to kindle their fires as their ancestors used to do. Why? Because the vines are so heavily treated that there is a serious risk of being poisoned. Burning a vine shoot is like burning an electricity pylon – its every last molecular fibre stuffed with highly dangerous synthetic substances.

As for arable farming, it has only recently been discovered that the latest types of high-yield seed, although they are more productive, are also more fragile than they used to be. Consequently they need not only more fertilising but more looking after to prevent them from succumbing to the slightest infection. Some of them need four times the amount of 'dope' to make them sit up and look pretty, at least for long enough to find their way into someone's shopping basket.

Marie-Paule Cépré, a reformed farmer who had spent thirty years polluting the countryside, put it succinctly: 'The more intensive the farming, the more pollution it will cause. In the pursuit of higher yields there will be overuse of fertilisers, chemical treatments and drainage. The humus layer is being eroded and fertiliser residue is penetrating the soil, reducing its natural fertility. The fruits of the earth are becoming unwholesome, a threat to human health.'[1]

Playing with Water

A courageous stand against a tidal wave. An immeasurable flood of fertilisers and pesticides has already deeply polluted the soil of France and northern Spain. And in the earth the liquid nitrogen turns into nitrates, whose ADI is 3.65mg per kilo. Not only that, but nitrates also turn out to be carcinogenic, and there are at least some in all the lakes and rivers of France.

In September 1992, some Breton farmers tried a surprising experiment: they attempted to drain the nitrates out of the soil using plants like spinach and beetroot which absorb them naturally. They withdrew up to 150kg per hectare! The vegetables were inedible, of course, but the lesson was ignored. In the past ten years, the annual

---

[1] From *'L'Année de la lune rousse'* by Marie-Paule Cépré and Danièle Lederman (Michel Lafon, 1993).

consumption of nitrogen-based fertilisers has increased from less than 1.5 million to more than 2.5 million tonnes. twenty-nine thousand tonnes of liquid nitrogen is excreted every day just by cows, calves, pigs and hens – that's three million tonnes a year, which is three times the amount of industrial and domestic waste. That too turns into nitrates when mixed with water.

According to experts, the soil is 'leached out' and the groundwater contaminated. It would have taken maybe ten years for the rain to wash the nitrates down to these underground water tables, and now they're working their way back up into our taps.

It will take at least a hundred years for them to die off.

Even the least polluted groundwater is far too polluted. Whereas European standards correspond to an ADI of 25mg of nitrates per litre, in France a level twice as high is accepted – a decision taken in 1994 by the court of Rennes, which thereby assumed a heavy responsibility. But it would have been somewhat unreasonable to make the entire population of Brittany do without running water.

Official predictions are rather depressing: by the year 2000, which is virtually tomorrow, there won't be a single water table still usable in the whole of France.

A Packet Full of Potions

It might be rather a sick joke, but how can you

...THE COLOUR OF FARMERS' BLOOD

not burst out laughing whenever you pass a field of sun-drenched corn being sprayed with huge jets of polluted water, knowing that there are scientists who advise us not to eat more than half a beetroot per month, to avoid lettuce, asparagus, cauliflower, spinach, artichokes and any other leafy vegetables, and to wash whatever we eat over and over again?

But washing fruit and vegetables no longer does any good. Nitrates and pesticides are unaffected, even by detergents!

ADIs also apply to fresh produce, but it is impossible to check them. Nor would anyone dare to attempt it. The Government has nightmares at the very thought of farmers up and down the country going on the rampage, abandoning their fields and descending on towns and villages in their tractors and manure carts to plough up the pavements, 'harrow' the forces of law and order or, worse still, 'disinfect' the streets.

And who would ever question the principle of intensive farming? Farmers earn the French government £25 billion a year in cash, which more or less covers the deficit in the social security bill and is twice the deficit of the Credit Lyonnais - often at the cost of their own and their families' health. The economics of the food industry outweigh all considerations of public well-being.

Unfortunately there is a lack of reliable statistics on the subject, but a quick dip into the closed world of cereal production allows us a fleeting glimpse of a terrifying reality: cereals cause serious illnesses more quickly and at an earlier stage than most other produce. It's not unusual for growers to die suddenly from cancer. There are many who suffer neurological trouble, and almost all of them develop crippling tendonitis, which creates a profitable industry for hypnotists and charlatans of every persuasion.

For fruit growers the problems are sterility and impotence, or their children might be born with abnormalities. In southern France doctors are only just beginning to realise how widespread the phenomenon is.

But since they avoid going to hospital, because they can't afford to leave their farms, these farmers are often unaware of the cause of their sickness. Most of them have never studied very closely the instructions and warnings which come with the chemicals they use year in year out. And even if they have, will they bother to dress up like astronauts every time they want to fertilise a field? You only have to take a trip through the countryside to realise that the answer is no. Most of them work in their shirtsleeves, without gloves or goggles. The lucky ones have tractors with enclosed cabs, but when the sun starts beating down on the roof, what can they do but open the windows? There are exceptions: the largest, but only the very largest concerns, like those to the south-west of Paris, where the rules are enforced and employees are obliged to wear protective clothing.

Alarming research carried out in the USA shows that more than 15 per cent of American farmers either can't read and write or aren't able to understand the instructions for the chemicals they use. What about French farmers? The Ministry of Agriculture refuses to answer that question, possibly because no proper study has ever been carried out. A pity, because people really do need educating about the dangers of farming. The use of pesticides has only just started out on its grand tour of the globe. Apart from Europe and North America, only Brazil, China and Japan have so far joined in.

DDT and Old Lace

Ronald Hites, from Indiana University, has found traces of dichlorodiphenyl-trichloroethane (commonly known as DDT), lindane, chloranil and eighteen other types of insecticide in the bark of trees growing in ninety different parts of the world.

Since DDT was banned almost twenty years ago in most industrialised countries (a ban unconditionally observed), the residue which is left on our plates must consist of molecules which have survived that long and crossed oceans without damage and without losing any of their power to harm.

In twenty years at the mercy of winds and currents, these molecules have had plenty of time to pollute both air and water before attaching themselves to some organism which they happen to have bumped into. Traces of chemicals used exclusively by developing countries in the Tropics have been found at the North Pole, thousands of kilometres away.

None the less, the pace of productivity continues to accelerate. But in a sense it isn't even off the starting grid and the engines of international speculation are only just warming up. In Chicago, where hundreds of millions of tonnes' worth of cereals are negotiated over 24 hours a day, 365 days a year, forecasters attempt to predict from one season to the next the planet's future demand for rice, maize, soya or wheat. Such high finance takes account of climatic, technological and political variables, but never questions of health.

Raking it In

In that sense 1995 was a pretty dramatic year.

> **Act I:** The scene is set two years earlier when the European Commission imposes a 'set-aside' (fallowing) policy on farmers in order to reduce production and maintain prices.

> **Act II:** A few weeks later, Russia, Argentina and the United States announce wheat harvests respectively 12, 18 and 6 per cent lower than normal. With total world production at barely 530 million tonnes, the lowest for twenty years, there is a shortage against demand of at least 10 million tonnes. As an immediate consequence, prices go through the roof in Chicago.

Act III: The EC does some sums and moderates its set-aside policy. French farmers produce between 65 and 70 hundredweight of wheat per hectare, sometimes as much as 90, compared with an average of 26 in the United States and no more than 15 in the former Soviet Union countries. If there's a market to be had out there, the EU can and will have its share.

Act IV: Back in Chicago, the speculators scratch their heads. Isn't there a danger of oversupply again, leading to another fall in prices? There are two critical factors:

– If China (the world's biggest wheat producer but also the world's biggest wheat importer) can't afford to meet its own needs, the other producers will lose a market worth $2 billion and 12 million tonnes of wheat will go back into the pool.

– If Russia is 2 million tonnes short (a chronic deficiency), how will the new Slav Mafias (who control all the markets in the former Soviet Union) pay for it? In real or imaginary dollars?

A total of 14 million tonnes of wheat threaten to drag world prices down. fourteen million tonnes which nobody will buy. Cuba and most of the African countries are broke. Elsewhere, the situation is complicated by local competition. Certain countries, even whole continents (Europe being a case in point) subsidise their exports for political as well as economic reasons. Others, like the USA, use their financial clout to influence the market. In 1994, despite being in the middle of a civil war, Algeria purchased 647,000 tonnes of wheat and 50,000 tonnes of flour. What with? You'd have to ask the American banks which handled the transactions!

Epilogue: In the administrative offices of an agricultural co-operative for the Chartres area, local cereal farmers ask the same questions as are asked in Brussels and Washington. They too follow the transactions as they happen in Chicago via computer screens. Should they invest thousands of pounds to achieve a record harvest which has little chance of finding a buyer?

It is the chemists who will decide: the manufacturers of fertilisers and pesticides have hundreds of credit arrangements on offer, with repayments spread over several years. Fixed prices, courtesy of the Common Agricultural Policy, will minimise the pain. Surplus produce will be stored in co-operative warehouses to be sold off at rock-bottom prices in the event of some disaster or other.

So what if there's also a surplus of pollution which will inevitably slip through the tiniest links in the food chain and end up on our plates?

5

SHOWDOWN AT THE OK CORRAL

Wherein we give you goose-flesh and rotten meat from cows fit for the funny farm, but it's no picnic.

Cereals have a surprising ability to concentrate all these chemicals and diseases before releasing them back into our bodies.

The most striking example are aflatoxins, toxins produced (as their name suggests) by the bacterium Aspergillus flavus. The bacterium develops in mildew which loves growing inside grain silos. Despite various treatments, it is found in bread, in oils and margarines produced from certain oleaginous seeds, in meat from animals fed on contaminated wheat or maize, in poultry, milk and eggs.

There's nothing innocent about aflatoxins: they have been used for centuries by witch-doctors in Guyana to poison people believed to have the 'evil eye', and are reckoned to cause a thousand times as many cases of cirrhosis and liver cancer in Africa, where storage methods are somewhat haphazard, as in the United States. All the same, between 30 and 50 per cent of American corn seeds (it varies from year to year) are contaminated by them. And since the USA are not only the biggest producers but also the biggest exporters of corn in the world, aflatoxins from across the Atlantic can be found throughout the food chain.

In this case, there is no 'official' ADI. Research carried out in Thailand, among a part of the population particularly prone to cancer of the liver, showed that sufferers had ingested a millionth of a gram of aflatoxin per kilo of body-weight per day for less than a month. Merely a quarter of the amount given to rats used as 'human substitutes'. There are no details as to where the toxic corn came from.

Nor do we know, due to the chronic paucity of French scientific research, how much aflatoxin is present in the meat of 'corn-fed' animals in France. And what a lot of them there are! 210 million chickens, 20 million cattle and 13 million pigs are 'grown' in France every year. Of course, production methods have had to be adapted, if only to be able to feed all these millions of tonnes of meat on two and four legs which are destined for our nourishment.

Once again, the worst toxins in the 11 million tonnes of fodder, cereals, flours and soya cakes (made from the remains of the plant after its oil has been extracted) sold annually to the breeders are concentrated little by little as they pass through the food chain. Not forgetting a further million tonnes a year (according to official statistics) of 'special ingredients', purchased on the black market at rock-bottom prices. Special ingredients bursting with chemicals (pesticides, insecticides, etc) declared unfit for consumption – even by animals.

But let's not forget either the most illegal substances of all, which are still often used to produce more meat more quickly: hormones.

Calving them Up

Officially, meat should no longer contain any hormones, especially veal, which was what caused the scandal in the first place.

In 1980 the French Consumers' Association, *Que Choisir ('Which?')*, campaigned for a veal boycott because the poor calves no longer even resembled animals. They were pumped so full of oestrogen (the notorious growth hormone) that they were nothing more than over-inflated skins which couldn't even support their own weight. Doctors were unequivocal: artificial hormone molecules were finding their way into the animals' tissue. Even today, when treating people with obesity, they make a detailed study of anyone who ate an excessive amount of veal as a child. It shows how strong these hormone products are.

On the other hand, no one has ever been able to calculate exactly how many people have developed cancer as a result of hormone substances. It's understandable because, like asbestos (an incombustible but highly dangerous mineral), they have a delayed reaction – sometimes as much as thirty years later. So it won't be until the next century that we'll get an accurate measure of our collective contamination.

By then it will be too late to find those responsible and exonerate them in time-honoured fashion. But already there are signs that those who dealt with these hormones on a regular basis twenty years ago are undergoing peculiar physical changes. Loss of body hair and the development of breasts are particularly alarming in a burly cattle-breeder. There are at least a hundred such cases across France, possibly more, but medical confidentiality is sometimes a useful smoke-screen when it comes to public health information.

Nevertheless, back in 1980, *Que Choisir* succeeded in arousing public opinion. So spectacularly and effectively, in fact, that within a few weeks sales of hormone-treated veal fell dramatically, first in France and then in the whole of Western Europe.

It was such a brilliant success that it provoked a sanguinary reaction from breeders: televised death threats, intimidating 'phone calls to *Que Choisir* reporters, herds of cattle used as roadblocks and coercion of MPs. As a result, the left-wing government's decision four years later (when Michel Rocard was Minister for Agriculture) to ban the use of hormones was a particularly brave one.

All the more so since the breeders continued to fly the flag of rebellion. There was even a threat from the extremely powerful (in terms of numbers of senators and deputies) National Cattle Federation: 'Hormone prohibition will merely encourage the black market, and specialist criminals, who are motivated by greed, will sell anything, including dangerous substances.'

Ten years later, when the use of hormones was banned throughout the EC by the European Commission, events unfortunately proved them right.

Multiplication Stables

In 'crooked' cow-sheds, hormones have been replaced by anabolic steroids, which 'transform nutrients into tissue' far more quickly than nature does. New chemicals like clenbuterol are produced in secret laboratories in South America, Miami or most recently the former Soviet Republics, where chemists will do anything to earn a few marks or dollars.

And the products are sold in more or less the same way as drugs: a secret rendezvous, a few grams changing hands, payment in cash. The prices are virtually identical too. One hundred pounds for a gram of steroids flogged by a Pole working undercover to a Frenchman in a clapped-out 2CV behind some disused village church. The same price as a gram of adulterated heroin sold by a Brazilian in drag in the toilets of a trendy Parisian bar. Surprising how much the town rat and the country rat have in common.

You could even say that the cattle-breeders are more 'with it' than their counterparts in the capital, since they have no hesitation, when the price of recognised steroids rises, in buying from labs which manufacture clones. Like computer chip clones, these are copies of the original molecules – just as effective but far more dangerous.

What's more, this steroid trafficking goes undetected not only by the drugs squad but by the veterinary services. Even if the animals are given steroids a few days before being taken to the abattoir, experts are unable to trace the chemicals in their bodies.

Meating Resistance

Occasionally a breeder is caught in the act and it sparks off a real street battle – or rather field battle. One such battle occurred in Belgium in early 1995. An inspector by the name of Karel Van Noppen decided to pay a breeder a surprise visit. Unbeknown to him, the breeder was part of a gang involved in an international meat racket. Taken by surprise, the crooks had no hesitation: an unmarked car ran the vet's into a ditch where he was finished off with a machine-gun.

Ever since then, Belgian vets make sure they have a police escort when they're visiting farms. That's if they can find any volunteers for such a delicate mission. Result: Belgian meat is more contaminated than that of any other European country – even Spain, where breeders also happily break the law.

Small local butchers, who buy their meat directly from an abattoir they deal with regularly, can always spot Belgian meat. After a week in the fridge (the usual maturation period before the meat goes on sale), the sides of beef turn an attractive blue-green colour and a sort of crust forms on the surface of the muscles as microorganisms multiply inside the steroid-inflated meat.

Don't they complain or demand a refund from the abattoir? No, because the animals have supposedly been examined by registered inspectors and all carry an official stamp authorising them to appear on our plates, either as mince or as steak.

There's nothing fictitious about all this. According to *Le Figaro*, a paper renowned for its seriousness, as many as 50 per cent of French cattle could also be affected. A proportion the breeders quoted in *Que Choisir* justify in the following terms: 'We're facing a crisis. For several years now we have witnessed unprecedented overproduction as a result of huge imports of meat from Eastern Europe and milk production quotas which have induced many dairy farmers to go into breeding. Add to that the constant fall in consumption [of red meat] in favour of poultry, and it's obvious that only the most competitive producers can hope to survive. In a situation where other European breeders are calmly seeing who can flout the most rules, we are left with only one alternative: do likewise or die...'

Or occasionally get caught.

The Steak Connection

In 1995 the Court of Poitiers sentenced twenty-one people including wholesalers, breeders, feed manufacturers, veterinary inspectors and butchers – a veritable 'Steak Connection' of white-collar traffickers who had been making profits estimated at millions. The same 200 in Cahors and in Chambery when two hundred breeders (at the last count) were summoned to appear before magistrates. On their farms, suits and ties were removed from their mothballs and even timetables for planes to Paraguay were checked. Just in case.

It's amazing that any arrests were made at all when you realise what a performance the 'Veterinary Service' inspectors have to go through to test for steroids.

Examining the animals for needle marks, however thoroughly, is generally a waste of time. The breeders are too cunning for that (the usual practice is to inject into the hooves) so the only practical method of detection is to take a urine sample.

This is a bit like a game of It's a Knockout. First the contestants need a bucket on the end of a long pole, a pair of wellies and a good deal of patience while they wait for one of dozens of calves, each one in a separate cage, to decide to take a leak. Then there's a mad rush toward the animal which is so terrified that it stops. This is the cue for the calf at the other end of the shed to start relieving itself and for the poor vets to race back again, slipping and sliding on the dung-covered straw, often with unfortunate consequences.

In the opinion of the defence lawyers, who are usually employed by the 'National Cattle Federation', 'the law is inadequate'. It's as if their ideal solution was a two-tier breeding system. On the one hand there would be battery calves allocated for three or four months to breeders on a fixed salary, who would have no option but to give them sackfuls of anabolic steroids and lorry-loads of 'protein enriched' feed in order to produce second-rate meat. This would be labelled as such and sold in hypermarkets and superstores. Poor man's meat.

On the other hand, and at considerably higher prices, there would be premier quality meat, specially selected and guaranteed hormone-free – 'Red Label' meat available at all good delicatessens. In other words, drugged meat for those of limited means, healthy for the wealthy.

And, undoubtedly, continued overproduction.

Oh for the Outdoor Life!

'What is the point in encouraging increased productivity when there are a million tonnes of meat piled up in fridges all over the EC?' This is the question being asked by the Farmers' Union, a minority voice whose members are more concerned with the moral and market values of the countryside than their 'industrialised' fellow meat producers. "The solution is to do the opposite and allow the small breeder to survive by encouraging quality produce. Consumers, who are becoming increasingly concerned with environmental issues and with the quality of their food, would certainly follow suit."

A round of applause for the speaker.

Another solution would be to change the way cattle are bred, to make it more humane. Once upon a time the cows were brought to the bulls, calves were given tender loving care, making sure they were being suckled properly, and the only limits of the pasture were those determined by the shepherd.

Such farms are a thing of the past. There aren't even many left like the one in Gueret, where the cattle spend the winter inside a nineteenth-century house, sleeping on the heated kitchen floor and helping themselves to hay under the vaulted ceilings or even upstairs, their muzzles resting on the wrought iron railings as they admire the sunset's reflections in the pond below.

At this farm even the cats can indulge themselves in a gentle massage from the local dog. Anyone who hasn't seen the toms sitting on the grass patiently waiting their turn as the previous customers depart with their fur slicked back with saliva, hasn't yet lived and knows nothing of the countryside they tell their children about.

Modern farming is more like mass production: the wholesalers and feed manufacturers are the factory managers and the breeders are the workers. The process starts with artificial insemination, the new-born calves are then taken and installed in batteries where they are subjected to a computer-controlled feed programme – an inexorable production line which gives the animals no respite and reduces them to meat on legs. Cows too old for calving are not even allowed a week in the field: they are sent straight to the slaughterhouse.

The calves are 'grown' in iron cages 1.7m long and 65cm wide where they remain for four months, their only distraction being the pleasure of drinking artificial milk from a rubber teat. They are taken to the abattoir at night, arriving (if all goes to plan) at first light to join thousands of other animals, all of them knowing they are about to die. They have to be given tranquillisers before the slaughterers can go to work.

Cowboy Practices

Consumers should not wait for the situation to improve. Once again, civilisation has given in to the profit motive. Just before world-wide deregulation (which was to come into effect on 1st January 1996), the Americans actually managed to have written into the *Codex Alimentarius* (the compendium of international food manufacturing standards) the concept of a 'maximum level of hormone residues' in meat. Which amounts to legalising the use of hormones and, *ipso facto*, insidiously sweeping aside European legislation. By the same token, the recent GATT could demand that the fifteen member countries lift their ban on anabolic steroids. We can look forward to some interesting transatlantic battles between diplomats, experts and consumers' associations.

This is because in the United States, for purely economic reasons, the use of hormones has become a way of life. They certainly don't have the same excuse as the Dutch, whose lack of pasture land means that their cows have to eat tarmac.

In the United States, as in France, a breeder can earn between £125 and £250 more per carcass if the meat has been treated and, as we know, public health counts for little compared with profits amounting to tens of millions of dollars. Sales of omatotropin alone (a powerful hormone banned until February 1994) have increased tenfold in the two years since it appeared on the market. It means that American doctors, following the example of their French counterparts, are now seeing a direct link between the rapid spread of obesity in the United States and the transfer of hormone residues in beef to human tissue.

Apart from that, Americans can't find words damning enough to express their opinion of European agriculture, which, according to them is in a disgracefully unhealthy state. Nolan Hartning, Emeritus Professor at the University of Iowa, reluctantly admits that 'Spain, Italy and Portugal do not meet our standards; as for France, she is occasionally admitted to our group of suppliers, but usually only with the benefit of doubt.'

Obviously no one has explained to this learned lecturer that the most noisomely poisonous chemicals are produced by laboratories which proudly display the star-spangled banner.

However, the Americans are not altogether wrong to be suspicious of European products, since hormones and carcinogenic steroids are not the only poisonous things in our steaks.

Cows that didn't Jump Over the Moon

The latest 'meat-madness' to hit the headlines is bovine spongiform encephalopathy or BSE, which everyone prefers to call 'mad cow disease'. Since its appearance in England in 1986 it has become the greatest fear, the worst nightmare, the *bête noire* of breeders everywhere. It is indeed an extremely serious disease, which can decimate entire herds. The cows become first agitated, then aggressive, they start crashing into things and stumbling over the smallest stones, and eventually they die.

The very first autopsies revealed that their brains had turned into a sort of sponge, with some parts dry and some wet, so that they were totally useless. After extensive research, enquiries, studies and surveillance, Her Majesty's Ministers of Health discovered the source of this terrifying epidemic: meal made from sheep bones, one of those 'special protein-enriched feeds' given to vegetarian cows all over Europe. That and other even more unmentionable substances like hydrolysed chicken feathers, braised blubber, various kinds of scraps, bad meat and hooves.

Nothing goes to waste.

It's true that the sheep whose bones were made into meal had a particular disease called 'the trembles', but the microbes which caused the disease would theoretically have been killed off if the meal had been heated to the regulation temperature. That didn't happen, possibly in order to save on gas. Whatever the reason, the British government, led by the 'iron lady' Margaret Thatcher, pretended nothing was wrong and went on exporting livestock to the continent without indicating that it might be contaminated.

It wasn't until three years later, in 1989, when a dozen British tomcats died after eating tins of contaminated meat from diseased animals, that the scandal broke and everyone suddenly woke up to the danger of mad cow disease. Experiments were carried out immediately: the trick with the killer cat food was tried out on a few more toms, which were then fed to some rats, which in turn were eaten by monkeys and so on. Each time, the spongiform encephalopathy was 'passed on'.

Rolling in the Isles

It was then that some of the more inquisitive doctors started looking at a peculiarly human disease whose origin was still officially unknown: Creutzfeldt-Jakob disease, known as CJD. This malady is characterised by psychological and psychomotor disturbances associated with the destruction of neurones. Although it can be fatal, it is most remarkable for remaining undetected in 90 per cent of cases because it is often confused with Alzheimer's disease, which also makes people go 'soft in the head'.

It was but a short step from there to the establishment of a direct link between BSE and CJD. The scientific world proceeded to don seven league boots and went in like paramedics called to the scene of a humanitarian disaster.

Because in the meantime another experiment had proved that a man with CJD could transmit the disease to an animal which eats his corpse. (In this particular experiment, human steaks were served up to a goat!) The difficulty of finding volunteers made it impossible to try the same experiment in reverse, so doctors could only assume (though fairly safely) that the opposite was also true. In other words, the disease could be transmitted from an animal to a human being. This meant immediate action stations in Brussels, hundreds of EC civil service commandos hurriedly air-dropped into British cattle-farms and a drastic decision: heads must roll.

It was now late 1989 and in a matter of days nearly 135,000 animals, a hundredth of the total cattle population of the British Isles, were to meet an untimely death. In addition, the export of any part of the carcass where the disease was most likely to be present (i.e. the bones, tendons and offal) was banned. All the remaining animals were to undergo rigorous medical examination.

But it was all too late.

During those three years, the diseased animals which were exported to all parts of Europe had time to mix with healthy animals. Dozens of cases of mad cow disease appeared in various places. How many exactly? We have absolutely no idea. The figures are 'top secret' and, in any case, bear no relation to reality. Firstly there are the animals known to have died before they reach the abattoir, and secondly there are those which have the disease but show no sign of it when they reach death row.

One thing is for certain: the epidemic is spreading. It was thought that young calves which had been raised on powdered milk and treated with antibiotics couldn't possibly be carriers. Wrong. Otherwise, why would the British government have decided early in 1995 to ban the domestic sale and consumption of veal offal from animals less than six months old. But only at home. They could continue to export baby calves – at a rate of up to 500,000 per year – provided they were still alive and could manage to stand up.

True British Phlegm

Anyone who saw the conditions under which these pathetic orphans were transported would have realised immediately that it was impossible, just by looking at them, to tell the difference between a calf which was dehydrated, distressed, panic-stricken, bruised and exhausted from the truck journey and one whose legs were shaking from spongiform encephalopathy.

Moreover, British exports naturally tended to increase with the fear of protest groups being formed at home. As a result, great convoys of martyred animals started crossing the Channel every day. The breeders had no time to lose. Animal rights groups went into action to defend the calves, but without success. Demonstrators were seen lying in the road while lorries drove over them. A woman of thirty was killed, live on television, but it had no effect: the scandal continued.

Mad cow disease poses exactly the same problems as hormones and steroids: the virus can't be detected and the resulting illness doesn't appear until many years later. They have another point in common: they are each governed by money. Truth hardly comes into it.

A confidential memo stolen at the time from an office at the European Commission in Brussels is quoted here verbatim: 'We must take a dispassionate view (of the mad cow disease situation) so as not to unsettle the market. We must no longer refer to the disease at all... The UK will be officially requested not to publish its research results... The problem must be minimised through disinformation. Our opinion should be that the media tend to exaggerate...'

Just like that.

All the same, the scandal really blew up at the beginning of 1996, with the UK counting her dead, the rest of Europe in a panic, frontiers closing and scientists arguing with each other across the barricades. It's a disaster not only for British breeders, but also for producers on the Continent, as beef sales slump everywhere. The agent responsible for transmitting BSE continues to spread, regardless of the blockade of the UK by the European Union. Entire herds are hastily slaughtered and incinerated to reassure the market, and while the British Government plans to massacre hundreds of thousands of animals, consumer associations demand to be told what it is that we are eating.

Firing Rubber Bullets

It is easier to understand now why honest breeders, particularly those who belong to the 'Farmers' Association', have been campaigning for years to enforce a system of 'tracing' meat from producer to consumer, so that butchers could reliably label it 'New Zealand lamb', 'Normandy beef', 'South Korean rabbit', etc. for the benefit of their customers.

In theory, such an 'identity card' for animals

already exists. Since November 1995 all cattle must be identified and the documentation must be accompanied by a colour-coded health certificate.

Green indicates that the animal comes from a herd which has been certified free of leukosis, brucellosis and tuberculosis. Yellow means that the livestock is regularly inspected. Pink shows that the animal is from an unlicensed herd.

The breeder apparently assumes liability when he signs and dates the documentation. It is further intended to impose stricter checks and higher penalties for contravening the regulations. All this should result in greater accountability and simplified export procedures.

On the face of it, some progress is being made. With the following provisos:

– The new health certificate is issued in advance on the basis of expected annual supply, which saves everyone a good deal of time (formerly, it was only obtainable at the time of sale).

– Other diseases legally identified as contagious, like hypodermasis, are not listed.

– A pink certificate means that the animal goes straight from the breeder to the slaughterhouse.

– The whole of Brittany is excluded from the arrangement for the next ten years.

Importers aren't exactly thrilled.

One wonders why. Perhaps because, being responsible for ensuring that the produce they bring into the country conforms to the regulations, they are now in the firing line in the event of a serious accident.

Ear Piercing Cows

Nevertheless, it would seem that the French are in the best position to organise 'Operation Identity Card' for animals. After all, it was a Frenchman, an engraver by the name of Chevillot, who at the turn of the century invented a method of marking the army's cattle using indelible violet ink so that they could be easily identified.

The system has been improved since then, because cows now wear plastic ear tags with a nine-figure number which is all that is needed to identify their country of origin, age and breeder. It's even possible to go one stage further thanks to another French invention, the 'smart-tag', which can carry a whole mass of information.

There's even money to be made from the idea, since the current market leaders, inheritors of the Chevillot company, tag some 200 million animals per year world-wide – a plastic earrings business with an annual turnover of more than £60 million. Which just goes to show that small businesses can be profitable.

But not all breeders are in favour of this system of 'animal tracing', especially those who have something to hide. And since they represent the majority, in France as elsewhere in Europe, it will be a long time before the butcher on the corner knows exactly where most of the animals he buys from the abattoir have come from. The consumer would therefore do well to cross his fingers before tucking into that steak.

Is this scare mongering? In March 1994, the German Minister of Health, Herr Horst Seehofer, put his foot in it when he said, 'We mustn't make the same mistake as we did with AIDS [sic] by underestimating the risks.'

At least sixty cases of mad cow disease have been diagnosed in Switzerland, which is not even a member of the European Community. And it has been discovered, sadly rather late in the day, that the incubation period of BSE can be as long as five years.

That of CJD varies between two and thirty-five.

6

THE POWER OF THE IMAGINATION

*There's everything in everything;
it's just a matter of finding it,
if not of warning the consumer.*

Let's just imagine for a moment that all the problems of hormone-injected meat and BSE are solved tomorrow with one wave of a magic sausage. Unfortunately, even if they were, our sirloins and our fillet steaks would still be full of rubbish as a result of all kinds of chemicals, some of which send a cold shiver down the spine.

This is the case with anti-bacterials, anti-parasitics and other antibiotics which are used in doses fit for half-tonne animals and which end up in the human body, even after severe grilling.

Anti-parasitics deserve a closer look. The advert for one of them, manufactured by a major American laboratory, boasts of its 'extended period of effectiveness' while at the same time warning not to administer it to calves, dairy cattle or pregnant cows.

The contraindications of other anti-parasitics claim that 'no ill-effects are known or have been observed to date'. Then immediately below, in red letters on a white background 'Danger. Bury containers deeply, and well away from drinking water supplies after use.'

And what about flumequine? This molecule needs no introduction to the men of the trade, it comprises 80 per cent of the anti-bacterial market and is adaptable to every kind of animal: calves, cows, lambs, pigs and poultry as well as rabbits and salmon and certain types of grain and cereals, fruit trees and vines.

As far as human consumption is concerned, *Vidal* (the pharmacist's bible) states in bold type: 'flumequine is not recommended for children under five'. Secondary symptoms might include giddy spells, blurred vision and dizziness. Not to mention related teratogenic effects, which have been proved in animals but are 'unknown in humans'. Teratogenic meaning 'creating malformations'.

Since a daily dose of 12 decigrams of flumequine can take up to a week to dissolve in the tissue, one can reasonably assume that in establishments where higher doses are used, the dissolution time will be proportionally shorter. Consequently, if this natural disintegration period is not observed, we will be unwittingly absorbing residues of the flumequine molecule.

And there's the rub, since no one knows the long-term consequences of these pharmaceutics, which vary according to whether you're eating Hungarian beef or Javanese rabbit and which interact over the years inside our intestines.

What we do know is that breeders use so many antibiotics as a preventative treatment that instead of being killed the viruses start to mutate. They are already able to resist even high doses of certain medicines. And once they have found their way into the human body, they don't even flinch when they happen to encounter one of the antibiotics we take – those tiny pills reassuringly prescribed by the doctor and measured to the nearest thousandth of a gram so that they don't knock us out.

But against these increasingly powerful invaders the only alternative is to use increasingly heavy artillery. It's what they call in geopolitics an 'escalating situation', the inevitable outcome being a nuclear offensive – in other words radiation treatment such as is carried out in cancer units (where gamma rays are a last ditch attempt to prevent the worst happening) and in food manufacturing plants.

Vacuum cooked meals, pre-packed salads, dehydrated vegetables and frogs' legs are the products most commonly 'irradiated' (i.e. radappertized or radurized), because they contain moulds, germs, bacteria, toxins, even parasites, which cannot be destroyed by anything other than X-rays.

Funny Young Birds

Meanwhile, in their own special death camps, millions of pullets rub shoulders in unrestricted intimacy while their breeders calculate the 'scientific' way of transporting them to the abattoir. They have to get there as soon as possible after the disintegration of the medications which they take like a drip-feed with their drinking water.

The breeders say they have no choice: if they wait too long, they'll have to give the entire batch another set of treatment – a two-week cycle which is rather expensive when the birds are only fattened for a few weeks until they reach the limit of two kilos, above which they are considered too big to fit inside a microwave.

Two hundred million chickens suffer that fate every year, not to mention several other indignities. There's no need to elaborate on the disgraceful conditions these poor animals are reared in; a mere glance at their feeding containers is enough to make your hair stand on end. The basic ingredient of their food, soya bean flour, has one disadvantage, a lack of amino acids, and in particular of methionine. The answer is to supply it artificially.

To make methionine you first have to convert petroleum into propylene – the stuff which exploded at the Los Alfaques campsite in Spain in 1978, killing 215 people. Then you convert the propylene into acrolein, an extremely toxic substance, then the acrolein into methionine and Bob's your uncle.

THE POWER OF THE IMAGINATION

All we need to know now is what percentage of each of these substances, from the radionuclides and the heavy metals through to the fertilisers, pesticides and medicines and not forgetting the hormones and anabolic steroids, remains in the animals' tissues when they arrive at the slaughterhouse. If there has been any research on this subject, it is kept quiet – very quiet.

Scheduled for Slaughter

A slaughterhouse is above all a chain of human activity which begins at dawn. The breeder has already done his bit: producing piglets ready to be weaned within a few weeks. To achieve that he puts the sows on iron beds inside cages where all they can do is lie there and suckle their offspring, who have to poke their heads through the bars. As soon as they are old enough to move onto solid food, they are taken away by lorry to be fattened up. After being force-fed on meal for six months (sometimes less), they are ready for the abattoir.

There we witness one of the biggest industrial processes of all time: killing and chopping up tens of thousands of animals every day and sending them off to be put in the fridge. A single abattoir in Brittany can kill 6,000 pigs a day, give or take a few. The official industry rate is 800 pigs an hour. A prodigious feat of modern ingenuity.

Every night, dozens of lorries full of live animals from all over the country converge on this 'factory'. The trailers are designed with the abattoir's production line in mind, and other similar trailers are already waiting at the other end to be loaded up with carcasses, blocks of frozen fat, vats full of intestines, hearts and livers, bundles of hides, etc.

Six thousand animals per day means 24,000 trotters, 12,000 eyeballs, 144 kilometres of intestines (enough guts in a year to encircle the globe) and 18,000 litres of blood. That makes 6,570,000 litres of blood a year.

That's just from one state-of-the-art abattoir. And if it's cattle rather than pigs, there's five times as much.

These quantities are checked minute by minute by the factory managers in order to keep production as close as possible to the capacity of the machinery. Blood, for example, requires its own special treatment: it must be prevented from coagulating before being sent off in sterilised tanks to its various destinations.

The blood is therefore poured straight from the animal into the mixers for defibrification. These machines run non-stop, twenty-four hours a day, their enormous paddles churning continuously the endless flow of lifeblood spurting from the dying beasts.

In such a factory a simple power cut is catastrophic: once the cooling process is interrupted, toxi-infections appear, serious diseases develop, germs proliferate and contamination is introduced.

Death is a promiscuous business, and the continual intermingling of animals of different races and origins on the same disassembly line – pigs from Belgium, Spain, Romania, Ireland – means a high risk of microbial contagion. A hundred thousand people are infected every year in France alone: dozens die of Listeria monocytogenes, brucellosis has yet to be eradicated, Campylobacter of the jejunum is rife, Francescella tularensis is a potential threat and the bacillus Clostridium perfringens is as much of a bugbear to scientists as bubonic plague, anthrax or tuberculosis. Not to mention the most splendid of all: faecal bacteria.

Let's start with Escherichia coli, known in the trade as 'E. coli', which develop in animal intestines and multiply as fast as cholera inside human stomachs. This is the notorious 'tummy bug' which holidaymakers get. Annoying and painful, it is more dangerous abroad than at home, because a healthy person's natural antibodies are a match for most native germs.

Unless they happen to be called salmonella or Listeria.

A Tocsin for a Toxin?

Lately, we have witnessed the triumphant return of the king of killers, Clostridium botulinum, the cause of botulism and ten thousand times as poisonous as strychnine. A scientist at the *INRA* has calculated that 'France could be wiped off the map with little more than 50g of Clostridium botulinum – roughly the same amount as ten cubes of sugar.'

Which is why it would be preferable if slaughterhouses didn't allow these bacteria to get totally out of hand. Since they exist even under normal conditions (every gram of our faeces contains several million E. coli), it's a case of 'the dosage defining the poison'.

Acute food poisoning is nevertheless rare in France, and will be even more so after 1996, when the country's 367 abattoirs have been brought in line with European health regulations. The safest slaughterhouses in the world, so they say.

The abattoirs of the twenty-first century will be more like space stations, with sterilised air, automatically disinfected corridors and supercooled chambers where slaughterers in pristine suits will slice up thousands of tonnes of meat with laser-cutters.

What will never change is the ultimate destination of the goods.

One Man's Meat...

Every day, our plates are crammed with more and more food made from animal products. Soon they'll be overflowing thanks to the work of the 'knackers' – the leftover merchants.

The leftovers mean the innards (heart, liver, kidneys, tongue, sweetbread, stomach), but also the lungs, spleen, bones, intestines, blood, feathers, hooves, skin and hide (when it isn't suitable for tanning). These have created an industry which employs tens of thousands of people and has an annual turnover of several hundred million pounds. Twenty years ago, no one could have imagined the significance it would have in our lives today.

Formerly the privilege of the abattoirs, the leftovers were the part of the animal the breeder wasn't paid for. Nowadays, they are bread and butter to the knackers, whose function as public servants is so highly regarded in France that they are granted a monopoly on the removal of dead animals. No small task when you think that in 1994 some two million of them were picked up: some died in the field, some in sheds or lorries, possibly victims of an overdose of steroids or even of spongiform encephalopathy.

Europe's first 'rendering' factory, located appropriately enough in Brittany, handles one and a half million tonnes of 'raw material' every year – in other words 'animal by-products' for which profitable, and often surprising, outlets have been found: varnish for tins, tile polishes, washing powders and fabric softeners, soaps, shampoos, waxes, candles, fertilisers (!), lipstick and make-up. Among other things.

Waste Not Want Not

Animal by-products are also a major ingredient of tinned cat and dog food. In fact it's one of their main uses, bearing in mind that there are more than 25 million pets in France, including 9 million dogs and 7 million cats (the UK has some 6.9 million dogs and 7 million cats), and the market is increasing by something approaching 20 per cent per annum.

The annual spend per animal is almost £250, which equates to 500,000 tonnes of food, 900 million cans and 600,000 tonnes of tin, making a turnover of £250 million, 50 per cent from exports.

The products of rendering have other, more bizarre uses. Cow's blood, for example, was used in the construction of France's biggest leisure park: Disneyland, near Paris, a fairyland whose walls are built not of concrete but of coagulated haemoglobin. It's a simple process: light polystyrene balls and cement are combined with container loads of blood, and the resulting mixture is sprayed under high pressure onto a fine mesh resting on a rigid framework, and this magical biological glue gives birth to a fairytale town which remains only to be painted in pastel colours.

But we might also find cow's blood in our farmhouse pâté, just as there is lard (another leftover) in rusks, cooked meats and ready-to-serve sauces. Pre-cooked chips will have been dipped in leftover beef dripping, and one would have to be extremely naïve to think that tinned duck conserve had been cooked in its own fat.

Grist to the Mill

Food producers are never short on imagination.

THE POWER OF THE IMAGINATION

Leading the way are the major grain harvesters (mostly American) who, as early as 1945, planned to enslave the world in order to sell their produce. Since America wasn't used as a battlefield during the Second World War, history decreed that she should be the world's granary. To this day, the USA's only real rival is France.

This explains why the august journal *World Grain*, the grain harvester's bible, is published in the United States (it retails at nearly £20,000!) and makes its very French counterpart, the *Catalogue officiel de l'agriculture*, look like a children's picture book. French harvests don't count for much against the US variety. The most conclusive proof is that two-thirds of French baguettes (and what could be more French than those?) are made from American, or even Canadian wheatflour: the grain being supposedly 'reinforced' or 'soil-enhancing' in contrast to French grain, whose deficiencies persist despite all the tonnes of chemicals poured over the country's soil each year.

The Americans study systematically every potential market. An internal memo published by Wheat Associates, the largest association of North American wheat producers, explains how to penetrate the South American market: by changing people's eating habits. One sentence reads: 'In some cases, for example in Colombia, traditional consumption patterns will have to be changed by substituting bread and pasta for corn and home grown potatoes.' Word for word.

To attain this goal, the United States adopt

their usual policy: free seed, guaranteed purchase, and bank loans. From Bogota to Valparaiso, the salesmen did their rounds until the local farmers, bound hand and foot by Wheat Associates, had no alternative but to plant their grain and increase the profits of the Association, whose sights were set on the furthest horizon.

They go prospecting in Africa, organise university courses in Japan and put money into promoting the sandwich in China. Wheat Associates even have the backing of the People's Republic to sell instant noodles to the Chinese.

But on the far side of the Atlantic overproduction is a constant threat, and animals in the industrialised countries are eating as much as their stomachs can digest. They already get through 90 per cent of the so-called secondary cereals: corn, barley, oats and sorghum. It requires a good deal of imagination to find an outlet for the remaining 10 per cent.

So whose crazy idea was it to feed humans like animals? Only the corn-flake manufacturers can answer that one, but these puffed-up grains of corn marketed with such vigour as 'good for the health' must be the biggest con-trick ever perpetrated on the human race. After all, a box of corn-flakes is nothing but a handful of popcorn – fresh air sold at the price of raw material. The packaging costs far more than the contents!

Nevertheless, this brilliant idea has gradually worked its way around the world, thereby assuring American grain producers a comfortable on-going income.

Which is the most important thing.

Soya Sauces

But did the grain men predict that their empire could be threatened by the cultivation of a leguminous plant with amazing properties of its own? The Chinese had known about soya for at least 3,000 years, but it was the Americans who tamed it and acclimatised it so well to the endless plains of the Midwest that they are now the world's biggest soya bean producers.

And these producers are clever enough to make the most of their resources, starting with animal feed, which is always a major preoccupation. Since soya is full of protein (40 per cent of its dry bulk), the tough, long-horned, buffalo which graze in the rocky mountains love it. It's their favourite food, and they pounce on the dehydrated soya cakes which are mixed in with their forage as soon as the lorries deliver it.

They aren't the only ones who love it: breeders world-wide are fighting over it. The French, who don't grow any soya themselves, import several million tonnes of it every year. In fact it's their third biggest import after petroleum and wood, enough to balance out the fantastic revenue farmers bring in to justify their subsidies.

And yet consumption has peaked. Despite world-wide soya-madness and annual production of 35 million tonnes, 13 million of which are exported, American growers are constantly in danger of over supply.

So they pay another visit to the Chinese (with 1.1 billion stomachs to fill, theirs isn't a market to be ignored) and try to start a craze for fish-farming using soya as food. The results are unimpressive.

Industry (that is, heavy industry) is a more promising avenue, and soya oil has already found a use in concrete. America's leading manufacturer of concrete slabs has recently converted its factory to produce soya-based lubricants for the moulds. Tests have shown that soya oil works just as well as petroleum-based products and is both cheaper and more environmentally-friendly. Since the concrete industry is currently paying between $3.3 and $7 per gallon for petroleum lubricants, it should be able to save several millions of dollars by switching to semi-refined soya oil which only costs between $2.8 and $3 per gallon.

The US Department of Agriculture is not short of ideas when it comes to helping its cereal producers. All its publications will from now on be printed in soya-based ink. A symbolic gesture? Not at all. The American Soybean Association has already worked it out: soya ink will generate $26 million a year.

But even this won't be enough to soak up the surplus. So the soya producers have come to a simple conclusion, just as the wheat and maize producers did: since animal consumption is no longer sufficient to exhaust the supply, soya will have to be incorporated into human food.

Return to Sender

Since a repeat of the corn-flake formula was out of the question (soya flakes taste awful, even with sugar), manufacturers looked for another recipe for success. One in particular had always made them lick their lips: soya steak.

Soya steak was the ultimate short-circuit, a way of breaking the cereal-animal-consumer cycle. By eliminating the meat stage, the cereal producers would cut out all the middle men who reduced their margins: breeders, fatteners, slaughterers and butchers, both large and small. This way they could deal directly with the supermarkets with the prospect of far greater profits.

The meat-eating consumer would of course have to be taken in, and that meant experimenting a bit. The French have experimented a good deal and have come up with a recipe for the future: mix a load of leftover meat with some soya, powdered egg-white, parsley, onions, a little xanthan (a sort of powdered latex) and lots of water, mix well and beat into shape and you will have a 'burger' which can be frozen and sold in supermarkets at unbeatable prices.

This isn't science-fiction. Soya steaks already exist. They're on sale everywhere. The only thing that distinguishes them from beef-burgers is the reference to 'vegetable matter' on the label. In very small print of course.

It's what people in the food industry call a 'new-product'.

7

YOU CAN'T MAKE AN OMELETTE WITHOUT BREAKING EGGS

Wherein we discover that in the modern kitchen it's the assistants that make the extra helping.

There really is nothing new under the sun, and when you look closely at the 'new products' (like corn-flakes) which manufacturers come up with, all they're doing is selling air and water.

Air costs nothing, and water, being naturally abundant, hardly any more. Water is irreducible and so provides bulk; heavy and therefore valuable; tasteless therefore anonymous; colourless and consequently able to adopt all the colours of the rainbow.

The only thing that is actually new is the way these elements are put together. 'New products' are simply food which has been taken apart, restructured, rearranged and glued back together again. Nouvelle cuisine only with bigger portions. Except that culinary experts don't talk about 'gluing' – that would be too vulgar; they prefer the word 'thickening' as in thickening a sauce. But the aim is the same: fluff it up into a mousse and sell air and water in place of the food itself.

This basic principle, when applied on an industrial scale, is taken to new creative heights with the help of chemistry. But one doesn't use the word chemistry either: one says 'technological assistance'.

It has become part of our daily lives. There isn't a slice of ham, a cream dessert, a tinned or pre-cooked meal which isn't 'technologically assisted' in some way or other. Manufacturers are particularly fond of technological assistance because it isn't subject to the restrictions of the acceptable daily intake. Their only guideline is the *quantum sufficit*, i.e. 'the quantity required to achieve the desired effect'.

And these are special effects – just like in the movies.

Deus ex Machina

Very special effects, indeed, when you see the list of most commonly used 'assistants': antifoaming agents, synergists, glazing agents, diluents, colour stabilisers, liquid freezants and anti-caking agents, humectants, enzyme, immobilisers, solvents, crystallising agents, flocculants, ion exchange resins, release agents and other lubricants, propellants and packaging gases, agents for controlling micro-organisms, leavening agents and additives to make things easy to wash or peel. Not forgetting those listed as 'miscellaneous' like anti-tartar agents, acidity correctors, firming and refining agents, and so on.

There are many others.

These technological assistants are not all ghastly inventions from the test-tubes of mad scientists. Some of them are quite natural. Everyone knows how to stiffen egg-whites, for example. So do manufacturers.

And they're not above doing so.

Sure as Eggs

World egg production is on the increase. In Spain alone there was an 11 per cent rise in 1994. Hens are 'encouraged' to lay more. Any farmer who keeps a few chickens for his own consumption will tell you that a good hen will lay between 60 and 180 eggs per year. But with the help of cross-breeding, artificial lighting and all sorts of other 'incentives', the big battery farms manage to obtain up to 280 eggs per bird per year.

At that rate it's not surprising the shells crack when you try to boil yourself some eggs. Older readers may remember the famous music-hall joke where a customer asks the farmer, "How much are your eggs?"

"A shilling a dozen if they're not broken, sixpence if they are."

Pause. "All right then, break me a dozen."

Those were the days, when the baker went

down to the local farm to buy some eggs and broke them himself to make cakes for Sunday afternoon tea. In the towns, horse-drawn carts would deliver eggs on a bed of straw. In winter, when the streets were icy, a sharp corner could mean a rather large omelette.

Then the factories took over and egg-breaking became a trade in itself. Highly-trained women using small knives could break several hundred eggs per hour for manufacturing spaghetti. Now they have been replaced by machines costing hundreds of thousands of pounds which can break ten eggs every second and separate the white from the yolk.

Indestructible machines which must make a profit. Machines which are responsible for producers demanding ever greater efforts of their hens, because the egg-breaking factory is always short of raw material. Ten eggs per second equates to two million eggs delivered every morning by articulated lorry.

To meet this demand France alone produces 15 billion eggs each year. It still isn't enough and they have to be imported from all over Europe, even though overproduction threatens this sector of the market as well.

These 'liquid assets' accumulate in super-cooled storerooms – yolks on one side, whites on the other – if they haven't already been exported in tubes or cylinders, or else dehydrated and powdered.

There's a constant need to create new markets to absorb the flood of merchandise.

Ovoproduction and Fishy Business

Like the Japanese, who are great egg-lovers, the Koreans buy them by the ship load. They use them for making 'crab-sticks': those seafood snacks they're so keen on. Fish is the basic ingredient, but the rest is all egg-white and water.

For the record, it was a French company which took the prize in 1994 for the exploitation of raw materials with fishless crab-sticks, made from a small amount of chicken breast dissolved in 60 per cent water with a few drops of egg white. Which goes to show that it's much easier to fish for chicken than for Mediterranean prawns.

Nevertheless, the cost of egg products is falling, much to the delight of food factory accountants. Eggs have caused a sensational breakthrough in 'performance' for the food industry. They are the technological assistants par excellence: whip them and they fluff up without complaining; flavour them and perfume them and you can incorporate them into almost any food, from processed meat to packet soup, from cakes to pre-cooked meals, from salad dressings to frozen pasta.

Eggs have become unbeatable.

Keeping us Sweet

The second technological assistant in today's cooking is sugar. France alone produces five million tonnes a year, and there's no escape from it.

It's a long time since manufacturers used powdered bone imported from Scotland to clarify the juice from the beet. Quicklime has replaced sheep skeletons and the beet growers are making a fortune as a result. It is rumoured that they make bigger profits than anyone else in the food industry.

There's sugar in everything. You won't find a pâté, a sausage or a packet of sauce that doesn't contain sugar as a preservative. An additional benefit is that sugar needs water to dissolve in – a useful factor when you want to add weight to food by soaking it before it's frozen.

The complete list of uses for sugar is instructive, but it would take another book or more to look at all of them. Practically everything which needs to be preserved for more than six months contains sugar. Advertising companies fight to get the sugar manufacturers' accounts. When you look down on the great European plains of sugar beet, like those in Normandy and Picardy, you might think the earth was a giant meringue.

Inscribed in gold letters on the front page of the sugar industry's instruction manual is the motto: 'If it's edible, there should be sugar in it; if there isn't, put some in.'

Diabetics beware.

Giving it Some Stick

The word carrageenan doesn't appear in many dictionaries, but it soon will. Whether it is extracted from seaweed or from exotic trees like guar, carob or xanthan, it always comes under the same headings: thickeners and gelling agents. In other words gums – vegetable gums like latex, from plantations where the sap is religiously collected between one monsoon and the next.

Religiously because it is extremely valuable to the Doctor Strangeloves of this world, concocting soya steaks in their famous laboratories. Valuable because it swells as it absorbs water and because the water it retains is undetectable. Theoretically.

It doesn't seem much, but when you squeeze a sachet of mayonnaise onto your hard-boiled egg in the canteen, it's gums (or carrageenan) which prevent it from dribbling pathetically onto your plate.

Ditto for the dressing on pre-packed salads. It's gum that makes it stick to the lettuce leaves. It's true: the world's greatest scientists work on gums with budgets allocated by the major food manufacturing groups, most of which incidentally are French.

Up a Gum Tree

The same research, using different methods, was carried out years ago by Mao Tse-tung. It had always been the dream of the great seers to feed humanity on gums. Mao Tse-tung had time on his side, as well as tens of millions of guinea-pigs at his disposal – all those who subjugated themselves to the Communist powers during the Hundred Flowers Campaign.

But Chairman Mao had quite a different problem: finding the Great Leap Forward which would put food in the mouths of hundreds of millions of people. And when the great famine occurred, city-dwellers died by the thousand while the peasants survived on grass.

It was time for some new ideas.

So the inmates of the 're-education camps' were given all sorts of experimental food, notably 'Triumphant Revolution' buns filled with pork made from a mixture of flour and sap – in other words gum. Being counter-revolutionaries and therefore 'viler than lecherous snakes', these unfortunate prisoners naturally had no more right to know what they were eating than they did to the little basil flavoured pork-balls which were supposedly inside the buns.

The results were on a scale appropriate to the eternal empire of China: in Cultural Revolution terms, immeasurable. After a few days of this diet of plastics, the weaker prisoners doubled up in agony, vomited blood and proceeded to join the Party's Roll of Honour.

The rest rushed to the lavatory in the hope that something would happen. But nothing did. The gums had swollen up and created a blockage which wouldn't unblock. They didn't all die but they all suffered, except those who were wise enough to spit the wretched buns out.

In the end, the number of dead was so monstrous that the experiment had to be cancelled for lack of volunteers. As one of the survivors said, 'It would have been less painful to die of hunger. There were so many corpses, the camp wardens didn't know what to do with them all.'

If at First You Don't Succeed...

In Europe the experiment was repeated, only with smaller doses of gum, proper medical and scientific supervision and rats instead of human guinea-pigs. What happened? If the rats weren't constipated they developed ulcers, with tumours of the colon as a bonus once the level of resins in their food exceeded 15 per cent.

Nevertheless, gums were declared 'fit for use' by every Health Minister in the Western world. Odourless, colourless and tasteless, they can be found hiding in almost every food we eat. They make yoghurts creamy and desserts glutinous, they add smoothness to packs of reconstituted ham, they form the basis of many sausages and the invisible structure of everything that's minced or reformed, they hold frozen pizzas together, allow cheeses to melt and so on. In fact, hardly a day passes without a so-called 'normal person' exceeding the 15 per cent gum limit.

And there's hardly a day when someone isn't admitted to hospital to have his digestive tubes cleared out.

For the manufacturers though, these gums which they use in ever-increasing quantities are the way forward. They make a mixture of water, egg, sugar and gum, then stir it up, heat it up and produce mashed ... something or other. Something or other which just needs defining, but that's the advertising agency's problem. What the manufacturers are concerned with is producing something edible at a fraction of the cost of real food.

'Now you know why manufacturers are constantly harassing us to replace the acceptable daily intake by the unlimited "technological dose",' explains an EC Technical Services official, adding, 'Fortunately for our future, they are nowhere near winning that particular battle.'

We shall see.

8

E IS FOR ADDITIVES

*In the struggle to please,
winning a housewife's favour
is a case of adding flavour,
and that is done with 'E's.*

Even if the proponents of the technological dose lose a battle, they won't have lost the war. And it's certainly a war we're talking about.

On one side, the front line consists of vast populations demanding sustenance, cowering behind the fragile barricade of democracy, precariously balanced on freedom of information.

On the other side of the firing line, the hordes of food manufacturers lay siege to them. Entrenched in the impregnable bunkers of trade secrets, they take careful aim at the bellies of the besieged, which are barely protected by the thin shield of the acceptable daily intake. Their officers have ordered them to use the most terrible weapons: additives. Meanwhile, from the crest of a hill, EC observers wearing blue caps embroidered with twelve gold stars watch the situation develop.

'Additives', it says in the official gibberish, 'is the term used for substances not normally consumed as foodstuffs per se nor generally present as typical ingredients of a foodstuff, whose addition in small quantities at whatsoever stage of the manufacturing process (manufacture, processing, storage, transportation, conservation or preparation) is intentional, be it for technical, organoleptic (affecting the sensory organs) or nutritional purposes insofar as it may entail alterations to the commodity.'

Such weapons include colourings, preservatives, antioxidants, emulsifiers, thickeners, gelling agents, emulsifying salts, stabilisers, flavour enhancers, acidifiers, acidity correctors, anti-caking and antifoaming agents, humectants, sequestrants, firming agents, modified starches, artificial sweeteners, bulking aids, raising agents, glazing agents and release agents, flour treatment agents, enzymes, gases and, of course, the inevitable 'miscellaneous' category (fizzing agents, colour stabilisers, etc.).

These are the famous E Numbers which litter the labels of packet food – not E for European but E for 'Evaluated', meaning in this case the evaluation of potentially harmful effects in relation to the acceptable daily intake. In other words, a sort of Geneva Convention for disputes between producers and consumers. No, there's nothing innocuous about E Numbers.

Would you like some examples? How about

E338, which is a preservative found in the world's most famous fizzy drink. According to the 'Institute of Physics and Chemistry' in Paris, it can be used either as a paint stripper or as a fertiliser. Take your pick.

According to the same institution, E320 and E321 are *persona non grata* in any laboratory because they are too dangerous. These code numbers conceal two of the most fearsome carcinogens: BHA, or butylated hydroxyanisole, and BHT, or butylated hydroxytoluene. These substances have existed since 1949, when they were first extracted from raw petroleum. It was Universal Oil Products which isolated the molecules right back at the beginning of the synthetic materials age. In-house engineers immediately realised that BHA and BHT would give plastics a longer life.

Plastics used in tins of army 'survival rations' which rapidly went into general consumer use, despite the fact that the original acceptable daily intake was set at a very low level, between 0.01 and 0.05mg per kilo, and has not been revised since. Yet, having fooled investigators for years, these diabolical molecules have finally disclosed their carcinogenic powers.

Colours to Taste

Ever since people started judging food by its colour rather than its taste, colourings became the manufacturer's principal sales tool.

The shop floor theory is simply: 'Colour determines taste'. And sure enough, human guinea-pigs used to test this theory, after having tasted and compared two identical products, one coloured and the other not, all give the same answer: they prefer the one which has been artificially coloured.

The same experiment can be repeated at any time in any shop or restaurant. Consumers will always go for food which is a warm colour, even in summer when the sun is shining and they're buying an ice-cream to cool themselves down. Red and yellow, the colours of meat and of the sun, sell well, whereas blue, green and orange are off-putting.

For that reason, no one would think of buying dark blue or fluorescent green fish, and yet these are the natural colours of several absolutely delicious tropical species.

If soya steak were presented in its natural colours, a greenish brown with pink flecks, the manufacturers would go bust. Since Western custom has it that raw meat should be bright red as if it is fresh from the abattoir, soya steaks are made red.

Making Oranges Orange

Fortunately for soya steak manufacturers (as well as distributors), red is the easiest colour to reproduce and there are innumerable ways of doing it. Beetroot juice is one. You only have to slice a cooked beetroot to appreciate how effective a dye it is. Beetroot juice stains, spreads and penetrates. There's only one drawback: being a natural substance, it oxidises when exposed to air, its bright red colour progressively turning pink before disappearing altogether.

So manufacturers looked for other types of colouring and called upon the services of a magic insect: the cochineal (from the Latin *coccineus*). Originating from South America, the cochineal swarmed across the warmer latitudes of Europe – Spain, Portugal, Italy and even southern France – where it is now bred by the million. Why? Because the cochineal is red – extremely red.

So red, in fact, that it was used in ancient times to dye cloth. The great Aztec priests made up their faces with it for funeral ceremonies, and in the Middle Ages, thread for the tapestries of 'Ladies with unicorns' was stained with cochineal. But our own civilisation has relegated the red insect to its proper place: the colouring of good red meat so that it practically sparkles in the neon glare of the superstore.

Cochineal comes in tins and is applied like a glaze, with a brush. Or it can be bought in powdered form and diluted in boiling water. Nevertheless, it's what is known as a natural colouring in the same way as carotenoic acids (carotene extracts) or chlorophylls.

On the other hand, there are chemically produced colourings like erythrosine (E127), tartrazine (E102) or citrus red, synthetic substances which are widely used because they last longer: they can survive long periods in fridges or freezers, on supermarket shelves or in tins. Erythrosine is the red colouring in many tinned sausages, and tartrazine (which has an acceptable daily intake of 7.5mg per kilo) is used to colour cheese rinds, the strange substance surrounding mass-produced pâtés, confectionery, desserts and above all, cakes.

As for citrus red, it colours things orange (!) including most of the oranges in our shops. In other parts of the world though, oranges are sold in their natural colour: green.

Those cherries in syrup, which look so healthy and appetising inside their glass jars, were never red to begin with. After being soaked in a cocktail of preservatives so that they can withstand years of exposure to light, they are dropped into a sulphite bath. Sulphites (derivatives of sulphur) like sulphur dioxide (E220) bleach the fruit before it is immersed in another bath of colouring. A nice bright red to make it all look the same.

It might amuse you to know that the use of sulphites (which have a particularly stringent ADI of 0.7mg per kilo) is totally illegal in the United States.

As with red, so with the other colours. Yellow is made from curcumin, lactoflavin, riboflavin and quinoline, or their chemical equivalents, and is used to colour butter and margarine, whose natural colour is white. Proper 'yellow butter' is the rancid variety beloved of the keepers of Himalayan yaks, but our own culture doesn't worry about little inconsistencies like that.

It's just as easy to make a mint cordial lurid green and so on. A mere glance at a 'natural exotic fruit juices' brochure is enough to show you the variety of artificial colours available: colours which even He who designed the rainbow couldn't have dreamed up.

Since the human eye can distinguish between 180 different colours of light, chemists armed with colorimeters – or, better still, spectrocolorimeters (which measure the diffraction of light in terms of its wavelength) – have unlimited scope for concocting ever-more appetising colours and ever-more sophisticated shades of red.

The Acid Test for French Dressing

Additives also come in manufactured forms such as are rarely found in nature. For example, in any cookery class you will learn how to make a French dressing. You mix some oil and some vinegar and you stir vigorously until they're more or less amalgamated as thousands of tiny bubbles. But as soon as you stop stirring, oil bubbles join up with other oil bubbles and vinegar bubbles join up with other vinegar bubbles until, within a few minutes, you're back to where you started. Unless you add some egg-white to trap the bubbles.

But since egg-white, like beetroot juice, is a 100 per cent natural substance with a tendency to decay rather rapidly, something better has been found: citric acid (E330).

Citric acid is a marvel of science. It is made by hydrolysing, fermenting and filtering maize starch, and 600,000 tonnes of it are produced world-wide every year, although that is almost certainly only the beginning. Citric acid disguises the taste of sugar and acts as a food preservative. Another benefit is that it's an emulsifier (in other words a fatty acid which absorbs liquid and therefore swells) – always an advantage when you're selling bottled dressing.

And the more the emulsifier is beaten, the more it swells and binds whatever is surrounding it: oil and vinegar, or oil and egg-yolk (to make mayonnaise), or water and powdered meat (which is excellent for pâtés), or water and flour (to make bread), etc.

Which goes to show just how magical citric acid is in terms of profitability.

All Roads Lead to Aroma

Of course, a dressing made of water and citric acid wouldn't taste of very much. No problem – a few flavourings will soon take care of that. Here, it's a case of making a silk purse out of a sow's ear.

The experiment with the colourings can be tried again, only using identical-looking foods with different fragrances. This time the idea is to get the consumer to say, 'It smells nice' or, 'It doesn't smell nice.'

In fact, with the help of additives, it's now pos-

E IS FOR ADDITIVES

sible to simulate almost any fragrance. How? With chemicals, chemicals and more chemicals. A fragrance is, after all, only an essential oil – a fact well known to perfume manufacturers, who mix them by the hundred.

It's more or less the same thing with chemicals. All you have to do is isolate the molecular structure in the animal's or plant's gene which corresponds to the particular fragrance and copy it. Then go into production.

The molecule governing the fragrance of an apple is one of the simplest to reconstruct because it gives off a strong, very distinct smell. Three drops of geranyl acetate are enough to create a whole batch of 'fresh fruit flavour' yoghurts.

It's the silk purse theory. But why not use real apples? 'Too expensive and too complicated,' chorus the manufacturers. 'Fresh fruit goes bad in transit, it needs storage space, the apples have to be peeled and sliced... It's a lot of work, you know.'

From the point of view of cost price, artificial aromas are always competitive. Natural vanilla fragrance costs approximately £600 per litre. The synthetic molecule comes in a hundred times cheaper. It's not exactly a photo finish.

The latest innovation to appear on the market is a collection of 'encapsulated' aromas which, instead of disappearing as soon as you open the tin or the sachet, are released progressively as you eat the contents. The tiny chemical capsules are dissolved by saliva, allowing both smell and taste to assail the palate so that every mouthful is a delight.

How do they work? It's ever so simple.

'Along the full length of the long linear amylose molecule,' explains Béatrice Condé-Petit of the Food Sciences Centre in Zurich, 'there are helical zones comprising more than a hundred glucose molecules. These zones make ideal traps for the aromatic components. In these helices the number of glucose molecules making up a complete ring, and therefore the diameter of the helix, is not constant. There can be six, seven or eight glucose molecules in a ring, depending on the size and shape of the aromatic component. For example, in the case of decanal, octanal, hexanal, ethanol and butanol, the helices are six molecules in diameter. Menthone, fenchone and limonene make seven-molecule helices, and eight-molecule helices are formed by naphtol."

Confused? You can make life much simpler by just using cyclodextrins, which now cost less than £6 per kilo.

This could well be the molecule hiding inside those amazing croissants, which you buy frozen but which smell just like a French *boulangerie* when you cook them. At first you are overcome by a warm aroma as if you were actually inside the shop – the smell that wafts into the street and brings the customers in. And the more you heat them, the more fragrant they become.

Then, when you bite into them, they give off such a 'home baked' smell that you might almost believe you were eating something fresh and healthy, forgetting that what you are busy dunking in your coffee is made from the discarded scrap of the food industry, and in particular from fat which for all anyone knows could be either animal or vegetable.

Grave New World

Researchers will literally go to the ends of the earth to find artificial aromas. In 1995, the *Revue de l'Industrie agro-alimentaire*[1] proudly announced that 'the Shinobu Gocho team from the Hasegawa Research Centre in Japan are putting the finishing touches to a process of bio-transformation of oleic acid in gamma dodecalactone.' (Sic) Or in the words of a specialist interpreter, 'The disintegration by micro-organisms of certain forms of hydrolised unsaturated fatty acids in C18 (calciferol) enables us to obtain a range of rare natural lactones which are extremely difficult to extract from vegetal sources.' (Sic again)

In common parlance, it means that the raspberry smell characteristic of wild fruit, which it hasn't been possible to extract by natural methods, will soon be obtainable by technological means.

Does anyone need a fried chicken and chips smell? Try hexanal 2.4 decadienal. A cooked meat smell? May we recommend dimethylhydroxyfuranone, first identified in 1976 in an old bottle of sake. (It's the same chemical that gives wine from

---

[1] Volume 537, 24th April-14th May.

the Jura region that 'characteristic burnt taste'.) This is what the cognoscenti call a 'neo-natural breakthrough'. Seriously.

Anyway, it's no laughing matter, because aromas are big business worth millions of pounds a year. And that's not to be sniffed at.

Glutamate and Tapioca

Aromas wouldn't be much good without 'flavour enhancers', the king of which is glutamate. No one has yet understood how it works, but it works pretty well; and once again it was the Chinese who invented it.

Of course, having had to survive countless famines, the Chinese have tried almost everything. They have only one rule, which is expressed by the proverb: 'Never eat an animal that walks on its back.' In other words, that's dead. But this hasn't stopped them creating hens' feet 'conserve' or jellyfish salad – two of the classic dishes of Chinese cuisine.

But the Middle Kingdom's greatest culinary invention is tapioca, a viscous, rather sticky starch which, when mixed with water, makes a perfectly satisfactory sauce. And it has the great advantage of absorbing not only the flavour but also the smell of whatever it's poured over, which is all the more impressive since tapioca itself has no particular taste.

All they had to do was extract the glutamate to create the king of additives: one which will retain water and absorb any flavour. Revealed here is a recipe for chipolatas as published by one of the trade journals: 75 per cent lean pork or beef, 12.5 per cent pork fat, 9 per cent water and 3.5 per cent tapioca.

And here's some of an editorial from the same magazine: 'Quite apart from the financial advantages of substituting water for meat, starch [i.e. tapioca and hence glutamate] allows a smoother texture to be obtained, reduces exudation and minimises loss during the cooking process.'

What more can we say, except one little thing: glutamate is known to cause neurological disorders. It's no coincidence that its effects are sometimes known as 'Chinese Restaurant syndrome', since there's so much glutamate in Chinese sauces that nausea and migraine attacks are included with the service charge.

Anyone who is allergic to white wine also knows what it's like. After all, people don't like seeing hundreds of tiny red spots break out all over their bodies making them itch in embarrassing places: a false allergy resembling urticaria (nettle rash) which is caused by the high level of sulphites in this type of wine.

True food allergies are much more dangerous. When they lead to anaphylactic shock or Quincke's oedema, they can even be fatal. People who are allergic to gluten, celery salt or peanuts know all about that. But these are ingredients which can hide behind references as deceptive as they are imprecise, like 'modified starch', 'spices' or 'vegetable fat', on the labels on the food we buy.

THE RUBBISH ON OUR PLATES

That's if there are any labels, which isn't the case in a restaurant or where caterers are about.

9

EXTRUSION-COOKING

Not to be confused with exclusive cooking, as in expensive restaurants, because there'll always be some rubbish left on your plate.

In the last ten years a miracle machine has appeared in food factories all over France. It is called the 'extrusion-cooker': the ultimate corn-flake maker, the father of the soya steak. No manufacturer can afford to be without one.

And it's so simple to operate: in one end you put a handful of grains of maize or wheat, and out of the other come dozens of boxes of corn-flakes. An extremely lucrative process with a record 2,000 per cent profit level.

But how does it work? It presses, compresses, cooks, beats, evaporates, stiffens, soaks, multiplies and amalgamates. The basic rule is always the same: use as little raw material as possible, that being the only element that really costs anything.

To produce a minced soya steak, you put into the extrusion-cooker some colouring (red), some water (as much as possible), some fat (animal or vegetable), some carrageenan (seaweed), some artificial flavourings (meat/spices/salt/pepper), a small amount of meat and, most important of all, soya. Stir it, deconstitute it, reconstitute it, put it on a tray, weigh it, wrap it and *bon appétit*.

You can also make mayonnaise in an extrusion-cooker. All you have to do is change the ingredients. A good mayonnaise requires powdered egg, water, sugar, anti-foaming agents (it wouldn't do if it exploded in the lorry), gums (to give it consistency), synergists, yellow colouring, colour stabilisers, liquid freezants, anti-caking agents, enzyme immobilisers, crystallisation modifiers, release agents, lubricants and finally, antimicrobial agents.

If there are any microbes left.

Meatless Meatballs

Even more spectacular is the meatball, which is essentially an 'emulsion' and obeys the basic principle that the more you beat it, the more bubbles it makes. Rather like soap, except that soap doesn't usually taste as good.

Meatballs are one of those regulated products which must conform to strict laws stipulating that the end product should contain a minimum of 13 per cent meat (fish, poultry, etc.). With an extrusion-cooker, it's child's play. First you take a little breast of chicken (preferably some scrawny old bird wasted from years of intensive egg-laying which has spent so long inside a rectangular metal box that it has ended up a similar shape itself), give the extrusion-cooker a tiny bit of that and a good dose of gums and press the button.

The machine will first slice the meat into minute strips and analyse its texture, then chop up the gums and beat them into the same consistency before mixing it all together. By the time the machine has finished, the chicken breast will have swollen by at least 30 per cent.

It's only the price of these machines which deters some manufacturers from getting them, but that won't be for long. The abattoirs have already realised how much they could make out of them, because it wouldn't be just one scrap of chicken breast which would expand by 30 per cent, it would be a whole lorry load – even a whole breed.

Thirty per cent extra imaginary chickens which won't have to be fed and watered and won't catch diseases. In a breed of six million hens, that means a gain of almost two million units. For the producer at one end of the chain as for the processor (sorry, the cook) at the other, that represents an awful lot of money.

Within a few days that same chicken breast has been dehydrated and packed up ready to be made into 'breaded turkey steak', 'chicken fillet in white sauce', 'braised breast of fowl' and so on: 'new products' to adorn the 'fresh food counters' of superstores everywhere.

Back at the meatball factory, there are things to be added to this plasticised chicken breast — water (as much as possible), flour, sugar, egg-white (optional), real or synthetic spices (usually synthetic) and some artificial colouring — before pouring the whole lot into a 'turbo-mixer' which can churn out a tonne of fully swollen product every twenty minutes.

It takes almost as long again to roll it into balls (well, that is what it's supposed to look like) and cook it. Once it has been boiled at 90-110°C, you can be pretty sure there won't be any bacteria left in it — not even dangerous ones. After all, these are the manufacturer's only real enemy — apart from the acceptable daily intake, of course.

At the end of the line, when the meatballs have been slotted into their tins and smothered with sauce, the only thing left to do is to stick the label on. There, under the small heading 'Ingredients', you'll find the gums (under the name 'carob and/or guar flour'), flavourings and 'aromatic components' and, as required by law, 'minimum meat content: 13 per cent'.

Then, in tiny letters at the bottom of the label: 'meat: 4.6 per cent of total net weight'. Which means that 95.4 per cent of a tin of chicken meatballs is things other than chicken.

That is the miracle of extrusion-cooking.

Creating Solutions

The miracle is due largely to the scientists and

EXTRUSION-COOKING

the R&D (research and development) departments of the big food factories. There, in the greatest secrecy, they experiment with different recipes, replacing soya by mashed peas, mixing starches and developing pre-cooked potato flour – very useful for packet soups.

It's also where they taste, or get people to taste, the various 'solutions'. The smell, appearance, taste and overall impression of each one, as well as how fatty it is, how moist or dry, how appealing to the consumer and how much it will cost, all this is noted, recorded and adapted (by colourings and flavourings) to the type of product envisaged, whether it's a pizza or a yoghurt.

And this is how they rediscovered one of the wonders of the food industry: processed meat.

Because the general public regards processed meat as the ideal 'preserve', it is infinitely flexible. It's all a question of imagination, chemical know-how and flair.

Hamming it Up

The chosen victim is, of course, ham, which is the number one seller of all processed meats. The poor ham is boned, chopped up, compressed and reconstituted to make it look vaguely like something of animal origin.

It's a long time since the days when farmers would soak a ham in brine so that it kept through the winter and then hang it up in the cellar, where the walls were grey with saltpetre.

Phosphates have replaced saltpetre, and the soaking process is now artificial: hams are no longer dipped in salt water, but stuck full of needles by a machine, each needle actually being a syringe. The object of the exercise, of course, is to inject into the ham as much water as possible (up to 12 per cent of its volume) to make it swell, as much sugar as possible to maintain the familiar pink colour and as many preservatives as possible.

The manufacturers had only one problem: whatever they did, as soon as they took the ham out of the freezer after a few days, water which hadn't been absorbed by the sugar ran all over the place. The ham leaked like an old goat-skin: hence the shiny appearance of those plastic packs of sliced ham. A repair kit was required, and for that the manufacturers turned to those famous exotic gums.

The latest recipe for ham is therefore as follows: having made a solution of polyphosphates, nitrites, gelling agents (usually from seaweed) and powdered proteins, add a few pinches of natural resins such as latex (also in powdered form) and cool the resulting liquid to an ideal temperature of between 0.5 and 2°C.

Inject quickly so that the resins don't block the needles and then kick the ham about to spread the mixture all the way through it. This is what's called the churning or 'massaging' stage, which in fact isn't done with a pair of boots any more, but in special machines which beat the hams for twelve hours at a time until they are sufficiently plasticised. The technical term is 'stabilised'.

Purists will also add carob flour, made from the carob tree obviously, a red-wooded, papilionaceous tree which grows in hot countries. Carob wood, as any cabinet maker will tell you, is an ideal hardwood for marquetry. What he won't know is that when its heart is transformed into flour by the foodstuffs industry, it makes 'an excellent gelling agent with a cohesive elastic texture', to use the official jargon.

But what does it do? It makes the meat easier to slice, which is absolutely essential because the ham has had such a pasting that it could hardly stand up if it weren't for the resins (which also retain the water). Besides, when you're making hams over a metre long (for institutional canteens) or big enough to fit exactly into automatic slicing machines, it's a good idea if the material actually holds together.

Separate Ways

In the same factories where the extra-long hams are produced, they also go in for horror productions like MSM – mechanically separated meat. The idea is to scrape every last remnant of meat off the animal's bones, be it the neck, the rib cage or wherever. This 'adherent meat' is a veritable gold mine for manufacturers, seeing that it accounts for 12 per cent of edible matter in a turkey and twice that much in a chicken.

Once inside the extruder, there's no longer any

distinction between meat and bone. The machine extrudes a red mush, which is immediately compressed into 10kg slabs 5.5cm thick, irradiated to kill off any microbes, frozen and sent off to other factories where they make little pots of baby food, dehydrated soup and, of course, ready-meals.

These slabs have a lot going for them. They're delicious mixed with dehydrated beef blood plasma, and even better with just a hint of meat juice vapour. And if you season them with a sliver of resin, a soupçon of seaweed, some colouring, flavouring, preservatives and stabilisers, it's virtually Cordon Bleu. Beat it all up one more time, extrude, reconstitute, coat with a semi-artificial red wine glutamate sauce, decorate with three slices of pesticide-treated carrots, add two potatoes impregnated with fertilisers and you have the very latest traditional style *Boeuf bourgignon*.

It's just as easy to make frankfurters. Allow 300g of 'MSM meat' per kilo of sausage, add some fat, approximately 300g of water, 200g of plasma, some artificial spices, garlic fragrance, nutmeg and coriander flavourings and some resin to hold it all together, and place in the extrusion-cooker at a low temperature.

Remove, prepare the skin (ex-sheep intestine modified by petroleum treatment), dye and leave to cool. The client prefers them smoked? No problem, pop them in the smoke bath. Would he like pine, beech, or oak? The choice is his. Nothing is impossible with a little high-tech gastronomic imagination.

There's no mention in all of this of what happens to the cooking water, but if it goes down the drain, you can be sure the fish are left open-mouthed. Unless (as is more likely) they end up swimming upside down.

Steep Bill of Health

The worst thing is that no one really knows what long-term effects all these weird and wonderful concoctions (apart from the known carcinogens) will have on the human body.

Doctors are not unaware that among the least understood elements in the development of cancer are nitrosamines. They have been recognised since 1956, but it's only now that they are causing concern. They are neither viruses nor bacteria, but simply a combination of amino acids (which are present in all living organisms) and nitrites, which are commonly added to cooked meats to prevent botulism.

So amino acids + nitrites = nitrosamines, and yet 75 per cent of the nitrosamines tested on laboratory animals have caused tumours in the liver, the pancreas, the tongue, the oesophagus, the stomach, the lungs, the kidneys and the bladder. Even with smaller doses, nine out of ten rats develop tumours in the kidneys and lungs within a year, 40 per cent of which spread to other parts of the body.

This is what Jean-Marie Bourre, Director of Research at *INSERM*, had to say in the columns of the monthly magazine *Que Choisir*: 'There are certain chemically synthesised molecules, like BHT and BHA (preservatives used in potato-based products), which are quite alien to our metabolism and therefore cause problems, although this has yet to be medically proved (...). At the same time, there are laboratories telling us to 'beware of the amounts used by many manufacturers because they can cause cancer'. It's total pandemonium.'

In the same publication Georges de Saint-Blanquat, then President of the Additives Group at the 'Public Hygiene Council', questioned the use of gums, which, 'although non-toxic, could cause physiological changes to our digestive system, particularly the intestinal tract, if consumed in large quantities.' (Remember the People's experiments of Mao Tse-tung.)

His conclusion: 'In any case, there's no point in trying to find out.'

Why not? Because it's too late. Except by putting the whole country to fire and the sword, it is impossible for us to escape the effects of chemistry, even though it is public health which is in the balance. But weighing heavily against it is the family food budget.

On the one hand there are a few thousand extra cases of cancer every year, and on the other thousands of employees and an annual turnover of tens of millions of pounds. It's hardly an even contest.

All Sweetness and Light

Barely a day passes without the government retreating before manufacturing advances.

> On Monday, they authorise the addition of 135mg of sulphur anhydride per kilo to biscuits containing less than 15 per cent fat.
>
> On Tuesday, they give bakers the green light to use 1 per cent of carob flour in their bun mixture.
>
> On Wednesday, all mustard manufacturers (except those of Dijon mustard) receive permission to increase the amount of xanthan gum in their products to 0.5g.
>
> On Thursday, there is official recognition of the effectiveness of a guar/xanthan gum combination (up to 2.5g per kilo) in 'restoring the elasticity and stability of egg-white after thermal treatment'.
>
> On Friday, seaweed concentrates force their way into prime quality ham.
>
> On Saturday, an artificial sweetener is added to ice-creams and sorbets in the proportion of 1g per litre.

And on Sunday, the kids fill themselves up with fizzy drinks (600mg of aspartame per litre), chocolate milk (ditto), cakes (1g per kilo), chewing-gum (5g per kilo) and jam (300mg per kilo), crashing through the barriers of the acceptable daily intake like a Formula 1 driver cornering too fast and finishing up in hospital.

The French obviously haven't read the Finnish research which denounces the aspartame molecule because it 'fosters premature degeneration: amnesia, dulling of the reflexes, uncontrollable shaking, etc.'. Is this why the Americans, who discovered the molecule in 1965, were not allowed to put it on the market until 1983?

There are those who maintain that aspartame should only be used in exceptional circumstances and under medical supervision. The French are happy to say that it shouldn't be given to children under three. But even this won't be for long; the ban will soon be lifted under pressure from the manufacturers, for whom aspartame is nothing less than a gold mine. When mixed with other artificial sweeteners, it is at least a hundred times sweeter than ordinary sugar.

Unless, of course, they simply ignore the ban, as they do with several other substances.

Earning a Crust

The 1994 report of the French Department of Consumption, Competition and Suppression of Fraud *(DGCCRF)* was devastating: more than 60 per cent of supermarket ham contains unauthorised additives or permitted additives in excessive quantities, 58 per cent of processed meats 'do not conform to regulations' and 30 per cent of cheeses contain 'illegal substances'.

EXTRUSION-COOKING

The authorities can't confiscate everything, so they turn a blind eye, as they do with daminozide, which is sprayed on fruit used in baby food. Millions of tiny pots are produced every day in the knowledge that as soon as daminozide is heated, it generates a carcinogenic molecule.

The 'race for progress' continues. Supermarkets are now being overrun by pizzas (natural flavourings, gelling agent E401, milk derivatives, modified starch, tomato powder, maltodextrin, thickening agent E466, soya-based cheese). The French bought 10,700 tonnes of them in 1994. And special types of bread are invading bakeries.

One recipe, for example:

Animal fat straight from the knacker's
+ lecithin E322
+ emulsifier E472(a, e & f) and assorted esters
+ antioxidant E300
+ mono- and di-glycerides of fatty acids
+ butylated hydroxyanisole.

Back-street shops are selling alarming mixtures of natural and hydrogenated vegetable oil or fat (copra, rape, palm, oil palm) hydrogenated animal oil (from fish), butter, water, glucose syrup, salt and emulsifiers (lechithin, mono- and di-glycerides of fatty acids and potassium sorbate). Not to mention acidity corrector (citric acid), beta-carotene colouring and diacetyltartaric acid flavouring. It's enough to make your hair stand on end.

With recipes like this it's possible to make 400-500,000 pancakes a day, which is 150 million pancakes a year. Some bakeries are producing up to 65,000 croissants an hour.

Indigestion

'New products' are everywhere: vacuum-packed sauerkraut, stew, kidneys, seafood dishes, cooked pork chops and sausages, gala pie, apéritif-flavoured meat cocktail, quiche Provençale-style, savoury loaf, mushroom vol-au-vents, stuffed pancakes, salmon and prawns en croûte, cooked spaghetti with its own sachet of cheese sauce (or beef or wild mushroom), ricotta cheese-filled pasta shapes, vegetarian ravioli, Chinese-style pancakes, battered prawns, coq au vin with 'new improved' sauce ('thicker, smoother, richer and shinier'), curried chicken, satay beef, etc.

There are others and there will be many more, because the food industry is always hard at work. The manufacturers' catalogues of basic ingredients are bursting at the seams with new concepts. Under W there's a proposed range of powdered seafood and seafood extracts; under X a duck concentrate which will allow products to be labelled 'with duck'; under Y a starch which dissolves in water and will increase bulk as well as crunchiness ('it won't absorb milk, so helps to keep cereals light and crispy in the mouth').

A is where you'll find texturising proteins to make things sliceable or spreadable; UCCS is a maize starch which is waxy at low temperatures to create soft, slightly spongy textures; Fla AFR73 is an extensive range of 'reconstitutable' cheeses in tubes... and Z is for a bacterial mixture of Staphylococci whose reductase nitrate activity stabilises colourings and encourages aroma production; and then there are microscopic algae grown in photoreactors, and so on.

In the 'semi-processed' products catalogues there are such amazing things as 'sake rice', seafood reconstituted as crunchy crusted flakes, an egg-white substitute, 'chicken broth extra' (one litre feeds a hundred hospital patients), prawn pulp (one kilo of pulp equivalent to two kilos of prawns), ready-to-use omelette mixtures, a type of cellulose which solidifies when heated but liquefies again when cooled, lecithins extracted from soya (half a per cent of lecithin achieves the same effect as eight times the amount of cocoa butter) for all kinds of cakes and sauces, J200232 for making court-bouillon, No. 200171 as a basis for onion soup, 1834 for a browned onion effect, etc.

Pull the other one!

10

SOWING AGAINST THE GRAIN

How Man the Creator admonishes God, whose work is found wanting in the eyes of nouvelle cuisine and its followers.

Like car makers and motorway builders, food manufacturers are beginning to realise that consumers are the goose that lays the golden egg, and that by killing them they are jeopardising their own business future. The retailers must make a crust, but so must their children.

But manufacturers are under no illusions: the concept of the acceptable daily intake will soon be a fact of life. Scientists are already blathering about individual 'counters', like Geiger counters only instead of measuring radioactive fallout, they will allow each of us to calculate our own contamination level.

Top secret research even suggests that customers would use these counters to read the bar-codes on each item as they went round the shop. If they did, they would undoubtedly go home empty-handed, because the needle would hit the red every time.

Bigger doses mean more cancers, more Alzheimer's disease, more CJD, more gastro-enteritis, more allergies and so on. The customer is left with a choice between dying of hunger and being poisoned to death. This isn't science-fiction; it's real life – minus the ADI counter.

So if the big bosses of the food industry want to survive too, there is an urgent need for them to reduce the doses of pesticides, insecticides, fertilisers, additives and 'technological assistants'. Although the Rubicon hasn't yet been crossed, it won't be easy to go back on the excesses of recent times.

It's pretty late already. The soil, the atmosphere, the oceans and rivers are, as we know, too badly degraded for us to hope that the planet will recover very quickly. But unless we want to stop procreating and abandon ourselves to collective extinction, like those prophets of doom who love to proclaim that 'the end of the world is nigh', we must do something. And not just anything!

Those who believe in progress through profit will tell us that science will take care of everything. Really?

Science Knows Best

What sort of science can come to our rescue this time? It's known as 'genetic manipulation' or 'biotechnology' or 'mutagenesis'. Let's call it ADD – the acceptable daily degree of this new horror – because 'mutagenesis' means 'gene transfer': extracting genes from living things, modifying them to achieve the desired effect and creating new species.

Since intensive farming is here to stay, it's a case of producing fruit, vegetables and cereals which will resist the onslaught of fertilisers and pesticides. Such varieties already exist: they're only waiting for permission to be put on the market.

Farmers will then go forth and multiply without restriction. Rape, beetroot, cabbage, potatoes and asparagus will no longer contain terrible and deadly chemicals which have already polluted the greater part of the northern hemisphere. Freed from the restraints of the acceptable daily intake, they will be able to increase yields (and profits) to their hearts' content.

But these 'manipulated' crops, these space-age seeds are terrifying Western scientists, who reckon that it won't be long before 'improved' genes find their way into the human body. Imagine it: the perfect consumer, a human mutant unwittingly programmed to feed at a particular time, on a particular day and in a particular place, responding automatically to the production schedules of the food industry.

When will this nightmare begin? Very soon if we're not careful: before the turn of the century. By the end of 1996 the Americans will already have put on the market a genetically modified maize seed resistant even to the maize moth, its most deadly attacker. When this starts laying into American maize seedlings, they will produce their own defensive toxins. At the same time, the growers reckon on saving a billion dollars a year on pesticides.

That's only the beginning. Mutagens of rape, beetroot, potato, soya beans and sunflowers have been developed and are being tested in the area around Arras and Metz. On the other side of the Atlantic, there's a genetically modified tomato, the 'Flavr Savr', which keeps for three times as long as ordinary tomatoes.

What's more, this tomato is a researcher's dream, because it meets the manufacturer's every requirement: it's full of essential natural vitamins, it's the ideal salad product (those tomato salads which conjure up images of hot summer days) and it can be pumped full of water.

It wouldn't take much to make it absolutely perfect. If only it stayed firm during shop opening hours; if only it jumped off the plant and into the farmer's pick-up all by itself, made itself exactly the right size to fit inside the packet, withstood a storage temperature of -18°C and arrived on the shelf without any unsightly marks or splits.

Well, it already ripens to order, so why not dream?

Peaches and Crime

The ultimate goal, the aim of every researcher, is

the variety with built-in pesticide, which will make all chemicals redundant. The question is: what will happen when the modified genes pass into wild plants and animals?

Because they inevitably will, just as they have already been transferred by the simple process of hybridisation. Nectarines are a famous example of genetic transfer, being a cross between peaches and apricots. Other types of peach are also alliances between different varieties which 'producers' blend over the years in order to obtain more and more uniform fruit: same size, same colour, same taste. It seems that uniformity is the key to consumer confidence, wholesaler satisfaction and retailer enthusiasm.

Since the French get through some 500,000 tonnes of peaches a year, the top of the range varieties are sold at exorbitant prices every winter by 'selectors', or are stolen from orchards on moonless nights. All you have to do is cut a branch off a 'modified' tree and graft it onto an ordinary plant to generate a whole crop of new mutagenic peaches and avoid paying 'royalties' to the inventor.

It's no joke: in the South of France, and particularly in the eastern Pyrenees, fruit growers have to employ private security guards to protect their trees from overnight attack by armed arboretum commandos. Overwhelmed by this new kind of piracy, all the police can do is seal off the trees.

While they decide what to do next.

Seek and Ye Shall Find

In other parts of the world, they're working on mutagenic bacteria to produce the chymosin or rennin enzyme – the active ingredient of rennet, which is extracted naturally from calves' stomachs and used in the manufacture of cheese. In the United States, they're training bacteria to produce artificial hormones which will make cows calve on a set date. The Australians have produced a synthesised molecule which causes Coquilles Saint-Jacques to 'retain' water. When frozen, they weigh 30 per cent more than normal.

But French researchers don't lag behind when it comes to increasing production rates. Especially those employed by the state to solve the problems of the foodstuffs manufacturers. They were given the following brain-teaser: given that a hen which has just laid will feel an irresistible urge to brood, and consequently no further desire to lay, how can the silly creature's maternal instinct be overcome?

The researchers put their heads together and discovered that the hen's pituitary gland secretes a luteotropic hormone, and the amount it produces increases in direct proportion to the number of eggs laid. So, after many and various experiments, with the international co-operation of McGill University in Montreal, these eggheads cracked the problem by manufacturing an 'anti-brooding' vaccine.

When the hens have laid their eggs, they no longer feel the need to sit on them and so rejoin the production line immediately. Productivity rules.

We can only hope that this vaccine has no long-term effects on the maternal instinct of our own nearest and dearest.

No Fun on the Farm

Poultry-farming has become more sophisticated than space exploration. You'd never guess how much cross-breeding has been done to arrive at a battery-hen which can manage to lay 280 eggs a year – twice as many as an ordinary hen. And in concentration-camp conditions: millions of birds with their beaks in a bowl (if they have any beaks left) and their backsides over a conveyor belt.

The birds least able to stand this horrifying prison regime are quails – sensitive little birds which jump at the slightest noise. Surrounded by thousands of their fellow creatures, they become stressed, shed their plumage, lose weight or turn to fat.

Not all of them, fortunately. A few, who are tougher than the others, survive without distress until they are sent to the slaughterhouse. How can you tell? By a simple experiment: put the quail on its back, with its feet in the air. Some will get up immediately while others will just lie there as if dead, their eyes shut in terror.

During a period of eight years, this test was repeated more than two 2,000 times in order to select the individuals with the strongest nerves.

Then, to complete the experiment, these game little creatures were placed on a conveyor belt in the middle of the battery.

The object of the exercise being to determine the birds' sociability – another prerequisite for collectivised overcrowding. Ones which had been free to roam would rejoin their friends within just 8 metres; those that had been crammed together would take up to 60 metres.

Amazing? No more so than growing plastic instead of extracting it from petroleum. Biologists at the Carnegie Institute in Washington have managed to isolate from bacteria a gene which determines the production of PHB (polyhydroxybutyrate), a biodegradable polymer that could be used for manufacturing, say, food packaging! When injected into watercress, the gene produces large quantities of PHB granules – up to 14 per cent of the plant's weight.

In another laboratory, bacteria have been trained to produce plastic film, which is used in the manufacture of Japanese loudspeakers.

Mutatis Mutandis

Scientists at the *INRA* (Institut National de Recherche Agronomique) have isolated chromosome No. 15 in the pig, otherwise known as the chromosome which initiates the deterioration of the flesh after the animal has been slaughtered. The transgenic pig, whose meat will keep for longer, is not far off.

And that's only the beginning, because in America they're working round the clock on grandiose schemes like the one for making goats produce human medicines. The technique is simple: remove a fertilised egg from a goat's uterus, inject into it the gene which will produce the desired molecule, replace the egg and wait five months (the goat's gestation period) until you have a calf whose milk will contain anti-coagulants or anti-inflammatories or anti-thrombosis molecules for myocardial infarction or molecules for treating emphysema or whatever you want.

Dozens and dozens of litres of it – a billion-dollar business.

And then there's Hermann, the famous transgenic Dutch bull with human genes so that the cows he inseminates produce milk suitable for babies. Or there are vaccines which automatically sterilise pigs. Of course, this kind of research is all top secret, because genetic transformation, like biotechnology, is kept under strict surveillance. Not unreasonably, as this little tale shows.

The University of Texas and the French Institute of Scientific Research for Development Through Co-operation[1] have announced the isolation of an extremely rare gene from an even rarer wild plant. This gene, which is known as apomixis (from the Greek *apo* meaning 'outside' and

---

[1] Formerly the Office of Overseas Scientific and Technical Research and therefore still referred to in France as ORSTOM.

mixis meaning 'union') has the unique property of determining asexual reproduction – i.e. reproduction without the transfer of DNA.

The plant actually reproduces itself, identically, so that it will remain unchanged until the end of time. And researchers have managed to transfer the apomixis gene to maize. They are congratulating themselves because peasants in poor countries will no longer have to buy grain, which is so expensive and which the industrialised countries guard like state secrets. Whether in Africa, parts of South America or Asia, farmers will be able to re-seed their fields year after year using their own plants, and they will always have exactly the same variety of maize. It's obviously a tremendous advance in the struggle against famine.

But what will happen when apomixis is transmitted to other plants and then to animals and then to human beings?

11

ALIMENTARY, MY DEAR WATSON

Or, when inspectors make their enquiries, they find some pretty rotten specimens.

In the end, we won't know whether we're eating real food, artificial food or semi-artificial food. Noëlle Lenoir, President of the advisory group on the ethics of biotechnology at the European Commission, has filed a report on this subject, a lengthy document in which she states: 'It is very difficult to assess these foods of the future objectively; even if absolute safety is impossible, the consumer of new types of biotechnologically produced food should not be put at risk. This is the No. 1 priority.'

The Commission has even enforced a particular type of labelling for 'genetically modified products'. But this little bit of sticky paper will hardly be an effective barrier against the invasion of

European markets by unscrupulous manufacturers. It's bad enough with the laws as they are, but when people start making up their own rules, it's beyond the limit.

There are importers doctoring, producers cheating, processors lying and retailers doing a bit of swindling too. Consumers are jumping out of the frying pan into the fire, and their protection associations, though well-intentioned, have little power in the face of such Wild West tactics.

An organisation which can make a stand against the French policy of laissez-faire is the Department of Consumption, Competition and Suppression of Fraud. In their brand new, ultra-modern building on the Boulevard Vincent-Auriol in southern Paris, the intrepid heroes of the *DGCCRF* dream up ever new ways of springing the con-men's traps. A stone's throw away, in an enormous steamship moored to the north bank of the Seine, the Ministry of Finance keeps a benevolent eye on its free-floating satellite.

But the con-men's imagination knows no bounds. Their inventiveness is almost as uninhibited when it comes to ways of cheating as it is in developing new technologies, and the free flow of goods created by the Single Market makes things even easier for them. To them, recycling or getting rid of damaged products from Denmark in Spain or from Greece in Ireland is a piece of cake. You don't need an agriculture degree to change the label and pretend it comes from somewhere else.

It's not only very easy, it's very profitable, and

the risks are negligible in comparison. A few months' prison at the worst. State controls have more holes in them than a fisherman's net. What's more, they have the annoying habit of letting the biggest fish (the sharks) escape and catching only the small fry.

Knowing how to navigate troubled waters is an art in itself!

Nothing New Under the Sun

'Plus ça change, plus c'est la même chose', as they say in France. If the customer is king, it's in the kingdom of the blind. He is so easily taken in that he hardly knows whether he's coming or going, let alone whether to say please or thank you.

Two thousand years before modern civilisation began, the consumers of ancient Babylon were so easily duped that the authorities introduced the Hammourabi Code to stop them being preyed upon. Part of this edict was specifically intended to control the sale of local beer. In those days, anyone found contravening the Code was sentenced to death.

Three and a half thousand years later, again in connection with beer, Erasmus warned against the overzealous denunciation of dubious commercial practices. He gave this example to prove his point: 'A preacher from Brabant was railing against those who deceive their customers by selling flat beer as if it were fresh, having added soap to it to make it frothy. On her return home, a beer merchant remarked that it was the most interesting speech she had ever heard. 'To think of all that money I've been losing', she said, 'by throwing away stale beer!'

Things hadn't changed much by the end of the nineteenth century, as witness the following excerpts from 'The Future Housewife', a volume intended for the edification of young ladies. It contains references to mixing chalk with flour, putting pieces of cloth and old leather in jam and making cocoa from 'lumps of clay with red lead for colouring and endive mixed with mahogany sawdust, powdered tanbark, red ochre and dried horse liver'.

You may smile, but such recipes are hardly less appetising than the chemical cocktails concocted by our own food industry.

But it's important to distinguish between what is legal and what is illegal. The dividing line is not always clear.

For example, in the extreme north of Italy, at the end of the Mont Blanc tunnel, is a French-speaking region, where hams and sausages are made by hand in the traditional way. In France, there is a village with the same name, where hams and sausages are made in a factory and sold to supermarkets. The latter are, of course, labelled in such a way as to make people think they come from Italy.

It's legal, so hard luck on whoever buys the wrong one.

More Mice than Cats

Talking of legality, science and technology aren't always on the wrong side of the line. Fortunately, the most sophisticated and up-to-date methods of investigation are also applied in defence of the public interest.

For instance, when it comes to analysing the true nature of a product, checking its state of health and verifying whether it conforms to standards, they make use of electronics, information technology, chemistry, physics and microbiology. They use all kinds of probes and measuring apparatus: viscometers, infra-red spectrometers, liquid or gas chromatographs, spectrofluorometers (which can distinguish between calf's meat and cow's meat), nuclear magnetic resonance and even carbon dating checks for wine. Not to mention the endless tests carried out by the 'National Centre for Veterinary and Alimentary Studies', the 'National Agronomic Institute', the 'Departmental Office of Public Sanitation' and goodness knows how many other public and private organisations.

But it's not enough.

Europe is only slowly coming together, so information doesn't circulate very quickly. In France it's even worse. It's not unusual for a spot check by one unit to undo all the determined detective work of another. Months of investigation and enquiry written off in a matter of seconds.

What causes an investigation to be made? Sometimes a complaint or an accusation or an indiscretion, but most often pure chance: 'random' checks, which nevertheless require a certain knack. Unless it's a case of responding to events: a sudden epidemic which has wiped out dozens of our four-legged friends, or an outbreak of salmonella or listeria poisoning which is keeping hospitals busy.

The Fraud Squad doesn't have enough staff to do any more. It's an unequal contest. In one corner there are 101 departments, eight laboratories and seven investigative units with 4,000 officers, not all of them in the field. In the opposite corner are the 4,200 companies comprising the food industry (400,000 workers), 150,000 retailers (1 million employees) and, of course, hundreds of thousands of farms.

The con-men are hidden amid the thousands of honest people, and those who are trying to catch them aren't even fighting at the same weight.

The Art of Persuasion

Another trump card in the hands of traffickers and law-breakers of every description is the existence of anonymous deals and co-operative 'arrangements' founded on a code of silence and the omnipotence of money.

The great debates among the business syndicates over unfair competition, which French breeders are supposedly facing due to the laxity of their European neighbours, make front-page news and give the politicians plenty to talk about.

A good excuse for doing nothing.

Similarly, when the Court of Coutances contended that 'the law relating to anabolic steroids had not always been scrupulously observed', it found that there were 'economic factors which help to explain if not to justify the motives of individuals who have infringed the regulations'.

It would be good to hear such words more often when magistrates are judging cases of shoplifting by victims of the market-economy – lest we should think it less of a crime to poison one's fellow human beings than to refuse to die of hunger!

Nothing to Declare

But let's not confuse the letter of the law with the spirit of the law.

Until recently, there were two kinds of anabolic steroids: those that were banned outright and those whose ban left room for manoeuvre. Whereas, for example, the prohibition of hormones was absolute, that of ß agonists was open to various interpretations. They were prohibited in animal feed but not animal feeding stations, nor when used for medicinal purposes. After all, these products had originally been developed for treating respiratory infections in animals.

In any case, there seems to be plenty of breathing space for the people who use them: ß agonists may not agonise beasts, but they certainly poison humans.

In Belgium and Holland, where land is too

scarce and therefore too costly to be given over to pasture, breeders make the best of it: local politicians more or less turn a blind eye, veterinary inspectors who look too closely risk the worst, and judges who are urged by their French counterparts to exercise greater vigilance suffer from occasional loss of vision.

The Spanish are no better. Yet over there, clenbuterol (a type of ß agonist) claimed 35 victims in 1990, 200 more in 1992 and another 136 early in 1994. The inevitable consequence was a couple of dozen similar cases in France. Borders only stop people; poisons travel freely.

By the way, when do we get a European *DGC-CRF* and harmonisation of controls?

Don't Worry, it's Nothing Serious

Here's another good one. Anyone who imports goods onto the French market is supposed to check that they conform to regulations.

The DGCCRF carried out a little survey to find out if this was actually happening. Unfortunately not. 'As a rule, importers merely ask their suppliers for samples, which they examine cursorily. When the actual goods arrive, they only carry out superficial checks'[1]. In fact, the majority of the twenty importers surveyed didn't even have proper testing facilities.

[1] The words of Jean-Max Charlery-Adele, from the DGC-CRF, as quoted in the *Revue de l'Industrie agro-alimentaire.*

So it's all a question of trust: importers must have unshakeable confidence not only in their suppliers, but also in themselves. After all, when it comes to food, the French are the best in the world, aren't they?

One might be tempted to think so. While in Germany, Denmark, Finland, Norway and Great Britain the hunt for the campylobacter had been on for some time, in France it was all quiet.

Radio silence remained in force until the results of tests carried out in November 1994 by the French Consumers' Association, *Que Choisir*, revealed that more than half the chickens examined were infected (over a quarter with salmonella bacteria).

No need to panic, though: if campylobacters really are the bacteria responsible for tens of thousands of food infections all over Europe, they're hardly a major hazard. They only cause a bit of a head-ache for a few hours or a few days, followed by loss of appetite, aching muscles and maybe a fever, and then some diarrhoea and severe abdominal pain for between two and ten days.

The simplest way to avoid gastro-enteritis is to make sure your chicken is thoroughly cooked. The same goes for pork .

Yet again, it has been proved that microorganisms, like radioactive clouds, don't give a fig for frontiers.

So what can the customs men do about it? For the most part, since the inauguration of the Single European Market, they have had to concentrate their forces along the borders with non-EU countries (like Switzerland) as well as inland. And they have other things on their plate (all sorts of taxes to collect, drug traffickers to catch, etc.) than checking the precise nature of goods going in and out.

Coffins on Wheels

Meanwhile, thousands and thousands of lorries travel backwards and forwards, transporting animals under appalling conditions, which have recently been condemned by various groups in Britain and by Brigitte Bardot in France.

Pigs, calves, cows, bulls, sheep, lambs and poultry, jolted about for hours and days on end as they travel from place to place or from country to country or from one end of Europe to the other, terrified, bruised, injured, collapsing in their efforts to escape, sometimes splitting their hooves and dying of thirst amid the stench.

Hell on earth.

When they get there, it's even worse. With a bit of luck, they can stretch their legs for a few hours (a couple of days at most) and take in some food and water to put on weight quickly before the final hammer blow. But the dead and the dying, which ought to have gone straight to the knacker, are often the first to end up on the hooks in the abattoir, despite their wounds and the incredible amount of toxins they have accumulated through stress.

The meat from these animals is like leather.

In the UK, animal rights activists have managed to get journey times limited to fifteen hours. But at the European Agricultural Summit on 22nd February 1995, ministers from the fifteen member states couldn't even agree on a limit which might be less restrictive. Nor could they agree whether lorries should be specially adapted for transporting animals, nor even whether veal cages ($1.5m^2$ pens in which calves are reared) should be banned.

A revealing anecdote sums up the problems inherent in such debates. When battery chicken farming was banned in Switzerland, the price of eggs (now laid by 'free-range' hens) went up so much that 'factory' eggs started being imported on a massive scale.

As ever, the consumer votes with his wallet.

The Garden of Eden Goes West

Although the *DGCCRF* experts can't be everywhere at once, they haven't exactly got their feet up. Some of them are checking the labels on dairy desserts with the billing 'fruits of the forest'. It's all a question of whether the fruit is wild or not, because, quite rightly, only wild fruit is allowed to be associated in the consumer's mind with a ramble through the woods.

But to find out if it really is, there's no point in opening the pot of yoghurt or soft cheese with the said label and tasting the contents. The surest method is to test it for pesticides, because if there are any, the fruit must be cultivated. That doesn't mean yoghurt can't be homogenised and 'fruits of the forest flavoured' by chucking in some pectin or carrageenan, for example. That's allowed, as long as the label says so.

No problem.

Back in November 1994 again, somewhere in Normandy, a whole stock of apples narrowly escapes mass destruction. After being kept for over a year in boxes inside a cold chamber, they have become contaminated by PCP. They'll have to be well washed to eliminate every last trace of this toxic fungicide, which is widely suspected to cause cancer.

In the same month, the journal *Que Choisir* warns consumers that pots of strained apple and prune baby food are so badly contaminated that a single gram of the mixture is enough to exceed the acceptable daily intake for a baby weighing 10kg.

Again in November '94, a report by the *DGC-CRF* on sweet ciders made from concentrated apple juice indicates that some of them contain patulin, an apparently harmless substance which develops in mould. But just in case, they set a limit of 50 micrograms of patulin per litre of cider.

A few days earlier, a new candidate for the Presidential election appeared on the horizon. His most widely quoted campaign slogan was to be: 'Apples are good for you.' It took a while to pick the pips out of that one, but that's neither here nor there.

ALIMENTARY, MY DEAR WATSON

Long gone are the days when you could bite into an apple called 'Striped Beefing', 'Beauty of Bath', 'American Mother' or 'Duchess's Favourite'. As late as the nineteenth century, there were nearly 3,000 different varieties in France alone, but by 1983, more than 93 per cent of French apple production came from just four types of North American and Australian apple. The famous 'Golden Delicious' had already swallowed up two-thirds of the market.

Apropos, it's worth mentioning how 'effective' gamma rays are in softening the flesh of the apple as well as in increasing its immunity to germs and accentuating the yellow of the skin.

Leaving something of a taste of paradise lost.

12

PUSHING THE TROLLEY OUT FOR SHEEP'S EYES

Wherein we look at the question of etiquette, since a brand can be just the ticket whereas a label leaves something to be desired.

Consumers are primarily concerned with saving money, but apart from that their greatest desire is to be reassured, flattered and made to feel important. Advertisers know this only too well and take care to butter them up. How? By making them believe that in choosing this product rather than that one they will demonstrate good taste, intelligence and an above-average lifestyle. They'll be happier too.

How many of us never fall for it? Everyone likes to be seduced now and again. But it's a pleasure which doesn't last, because fine promises are seldom kept and the trick is to make you forget that in a 'sparrow and horse' pâté there's a whole horse for each sparrow – and a Trojan horse at that.

If we only got caught once a year or less, it wouldn't be too bad. But under constant and insidious pressure from advertisers, consumers surrender to every passing fashion, whether it's 'ready to eat' or 'light' or 'simple' or 'natural' or 'real' or 'traditional'. And when they come up with something 'genuinely freshly organically authentically ethnically instantly eatable', don't stand in the aisles or you'll be mown down in the rush.

Especially if there's a brand name to go with it, a name branded onto your retina by endless TV commercials. They wouldn't spend so much money moulding plastic cabbages which won't wilt under the studio lighting or painting plaster pies for magazine covers if it wasn't worth it – i.e. if they didn't sell.

Seeing Red

And they sell almost as well as 'Red Label' products: this is the distinguishing mark of 'superior' French produce, which is awarded in accordance with guidelines set down by an ad hoc committee advised by the Ministries of Consumption and Agriculture. To earn it, a producer must demonstrate and provide long-term guarantees for the unrivalled quality of his merchandise.

There are no tricks here. Even the consumers' associations all recognise the genuine superiority of, for example, Red Label chickens. They have much better food, consisting of selected meal and seeds, they can stretch their legs and get the occasional breath of fresh air, they are less crowded and, finally, they are allowed to live at least eighty-one days before being taken to the abattoir.

There are even those who are so obsessed with hygiene that they wait a short period before putting in the new batch of chickens. As a result, they produce much better meat: it's firmer, tastier and more fragrant, besides losing less fluid when you cook it. The *nec plus ultra* of chicken. Even on the salmonella front, only 8 per cent of Red Label birds carry it, compared with 34 per cent of battery-chickens.

Who cares if, for some unknown reason, two-thirds of them contain campylobacters as against just over half the standard variety[1]? It's far less serious and detracts nothing from their superior quality.

There are always two ways of looking at things. You can claim that Red Label chicken is far better than its rivals, or you can simply acknowledge that it is not nearly as bad as normal chicken.

But to really understand the difference between natural and factory breeding, Red Label or otherwise, you have to taste a chicken which has been reared in total freedom, living its life out in the open pecking about the farmyard. True gastronomic pleasure is always to be found at the breeder's own table.

What applies to chicken also applies to pork: try one of those slightly greyish chops instead of

---

[1] Figures taken from the magazine *Que Choisir*.

the artificial pink ones you normally consume.

Anything that has been naturally raised or grown is naturally better. They're not stupid, these breeders and farmers and foodstuffs manufacturers; they know what's good for them.

They leave the rubbish for everyone else.

Organic Panic

No more fertilisers or chemically synthesised pesticides and no more nitrites or nitrates? That's a good one!

Could it be possible to eat without risking being poisoned, to go shopping without having to calculate acceptable daily intakes or even to give yourself occasional indigestion from having rediscovered the pleasure of taste?

It could just about happen, so let's admit it straight away. But only products bearing the AB[1] logo, which is awarded by the Ministry of Agriculture, are guaranteed to be 'organically grown', or at least to contain 80 per cent organic ingredients.

> Question No. 1: Why can't they contain 100 per cent organic ingredients?

> Question No. 2: How can 'organic' products be given pure (i.e. nitrite-free) water and be grown or cultivated without being affected by acid rain, atmospheric pollution or radioactive clouds?

---

[1] Which stands for agriculture biologique.

The obvious answer to both questions is, 'They can't', but that's easy to say.

For the record, in the summer of 1995, pots of a certain 'organic' yoghurt suddenly sprouted an extra label, which said: 'This product is not organically produced.' That was easy to print.

Thanks to the *DGCCRF*.

Making Mountains out of Molehills

The more controls there are on branding, the more the big brands try to bamboozle us. And when it comes to bamboozling, there's no limit to the ingenuity of marketing departments. More and more wines and beers, cheeses and milk products, meat and poultry, 'specialities' of all sorts and even certain fruits (prunes, for example) are marketed with specific regional or local names, but these may or may not be authentic and bear no relation to the quality of the product.

The towns and villages might do well out of it, but not necessarily the consumers. Nor even the manufacturers. That so-called 'traditional farm lamb', for example, promoted so energetically through televised commercials, is trespassing on breeders who have spent years trying bravely to establish the superior quality of their produce.

And then there are the local rivalries. Unfortunately, when it comes to geography, the food industry follows its own peculiar logic. The use of the description 'mountain', for example, has nothing to do with altitude — in fact, far from it. It ultimately depends on the goodwill of the local official, who apparently insists on certain guarantees and conditions. What's the betting that his perspective bears little relation to the geographical contour lines?

Even in the European capital, Brussels, they seem to know something about mountaineering. In fact, they have only recently scaled new heights. On the recommendation of the European Commission, the Environment, Public Health and Consumer Protection Committee of the Strasbourg-Brussels Parliament has introduced a new rule for the protection of 'traditional national products'. It states that any product labelled 'traditional' shall not contain any additives. None at all!

That's certainly aiming high.

So they invited each member country to submit a list of products for consideration. And out of a total of 300 proposals, they accepted fewer than a dozen. France had the top score with four: bread, tinned truffles, snails and conserves of goose, duck and turkey. The Germans planted the flag with their beer. The Greeks came up with feta cheese (made from ewe's milk), the Austrians with *Bergkäse* ('mountain cheese'), the Finns with *mämmi* (a dessert), the Swedes with a fruit syrup and the Danes with meatballs and goose pâté.

We'll soon see this new 'brand' appearing on labels. But although the products displaying it will offer a certain guarantee of quality, it is likely to be short-lived. While all this has been going on, a 'green paper' on food laws has been fermenting in the cellars of the European Commission which proposes to abolish 'excessive legislation imposing a restrictive burden on the industry' on the pretext that it is up to consumers to read the information on what they are consuming.

Touché!

Busting a Gut

The obsession with 'low-fat' products is fortunately a thing of the past. Everyone has now realised that they were 'invented by manufacturers wanting to increase their prices', as the Research Centre for the Study and Observation of Living Standards[1] so pertinently put it.

In fact, a low-fat diet is the best way to get fat, as was proved by an American survey of 80,000 women aged between fifty and sixty-nine. The body is only fooled for a while – hunger pangs soon return and the missing calories are quickly added back in.

So much for that ideal.

The 'light' (i.e. 'sugar-free') craze was the next one, helped by artificial sweeteners like aspartame, although it would be better to replace quick-acting sugars by slow-acting sugars than to stimulate an unnatural craving.

That leaves 'slimming', 'diet' and 'fitness' foods with added vitamins, minerals and trace elements. Here again, it ought to be safe to say that,

[1] *CREDOC.*

although they might not do any good, they don't do any harm. The problem is the famous 'guaranteed minimum content', which attracts vitamin addicts like moths to a candle.

Because the contents are far from being guaranteed. In more than a third of cases, the quantities specified on the label are not adhered to; some products have been found to contain unauthorised ingredients; the amount of vitamin B1 in fitness foods exceeds the legal maximum; and food supplements go through the roof when it comes to the recommended daily intake.

Guess who found all that out...

Nutritional Claims

...The *DGCCRF*, of course – the inescapable *DGCCRF*. It quite often trips the industry up, but of course everyone knows that the industry is always two laps ahead and has more than one runner in the race.

Having extolled the slimming potential and the supposedly guaranteed vitamin content of their products, manufacturers 'quite naturally' get around to trying to persuade us that by eating them we can improve our health or cure ourselves of illnesses. But such 'therapeutic claims' have been anticipated and strictly forbidden by European directive. After all, refrigerators aren't medicine cupboards or vice versa. Otherwise, every meal would be subsidised by Social Security.

But European directive or no European directive, advertising agencies have a knack of playing with words. Or should we say, for making suggestions as opposed to explicit claims. But watch out – the *DGCCRF* watchdogs know a thing or two about ambiguity. Here's one slogan that didn't get past them: 'Your body has its own natural defences. X can help you strengthen them.' In this case, X was a well-known brand of yoghurt.

An even more devious method of getting round the law and avoiding the regulations relating to the promotion of slimming products is to use members of the public as your sales reps. Spreading the word from the comfort of their own homes with nothing to intervene between them and the customer, these minions acting on behalf of some nebulous multi-national can get away with murder. They can say anything they like. How about 'You can cure your whatever just by eating such and such once a day. I know, because I did it...'?

Even the *DGCCRF* can't do anything about that.

13

IN VINO VERITAS

*With our taste buds in queer street,
we hit the bottlenecks of Paris
and find ourselves in a jam.*

They say that there is a wine for every dish. So let's be totally logical and completely intemperate and wash down our bull's blood with a glass of noble rot. We could even add a drop of red to the poisoned chalice.

It's because they didn't want to be in the red that wine makers learnt how to get the most out of their vines. Even in Roman times they knew how to do it, but the patricians soon had them pull up all the vines which produced too much. They weren't stupid, those Romans! When the Gauls inherited the Roman vineyards, they gave up their barley ale and took viticulture to new levels, sometimes achieving more than 8,000 litres per hectare.

Of course, quantity and quality rarely go together and such practices don't make good wine – you don't have to be an oenologist or a wine connoisseur to realise that. Even the least knowledgeable of wine drinkers notices it and eventually turns away from French to, for example, Californian, Australian or Chilean wines, which are cheaper into the bargain.

Vile Vines

How could wine makers have sunk so low? The answer is quite simple: in the same way as the grain growers, who are now producing more than 80 hundredweight per hectare. Vine growers' profits increased by nearly 34 per cent in 1994: that means they produced 34 per cent more grapes.

They start by selecting the most productive vines: plants which have been 'cloned' in nurseries, i.e. reproduced identically ad infinitum from the same stock. But uniformity has its down side: a lack of complex aromas and perfumes which make all the difference between a good wine and an ordinary everyday plonk. Which is why a bit of clever mixing goes on between different stocks and different vineyards to try to put some taste back into the wine.

The selected vines are first of all given the chemical spraying treatment: the maximum dose of manure, herbicides and plant disease treatments. Whereas thirty years ago less than 15 per cent of French vineyards underwent treatment (in those days with simazine), in 1995 85 per cent were subjected to chemical weed-killers.

In less than a generation, growers had to develop all sorts of additional treatments to combat weeds (like amaranth and senecio, etc.) which have sprung up or become resistant to punishment. Then it was time for the cocktails, great mixtures of sulphites and nitrites, combinations of substances which were increasingly virulent, increasingly difficult to handle and increasingly dangerous for the consumer.

All without significant results: as the treatments have become less and less effective, growers have reached the stage where they now alternate between residuary and foliar products.

We should mention in passing that the so-called 'residuary' herbicides like diuron dissolve in water. As for the weed-killer simazine, whose use has been restricted since 1990 to 1.5kg per hectare, this was classified as a 'possible carcinogen' ten years earlier by the European Commission, which set the acceptable dose in drinking water at 0.1 microgram per litre (a limit which has been raised to 17 micrograms per litre by the World Health Organisation and is still exceeded in many cases). At the Conference on North Sea Protection it was even recommended that its use be 'restricted'.

Add to that the 'minimum shoot size' (they are pruned as little as possible to ensure the maximum quantity of grapes) and yields are bound to go through the roof.

The Rot Sets In

But why stop there? Another common practice is early harvesting, so as not to take too many risks with the vagaries of the weather (hail storms, frosts, etc.). The result? Just as corn is no longer suitable for making bread as it is, grapes aren't ripe enough for making wine, which comes out weak and watery. But don't worry, there are plenty of ways of sorting that out...

Once the grapes have been picked and pressed, the fresh juice is usually left to ferment at a low temperature, with a bit of artificial yeast and some sugar added (chaptalization) to raise the alcohol level.

Chaptalization is strictly controlled. It is only supposed to be done when the grapes haven't had enough sunshine, but in practice permits are handed out pretty generously by professionals within the industry – men of taste, of course.

Then there's the fining and filtering stage, modestly referred to as a process of 'clarification'. The idea is to coagulate the liquid and force suspended particles either upwards or downwards. The most effective methods involve adding bull's blood, fish meal, gelatine or casein – whichever. Fining and filtering see to it that there's no tannin left in the wine by the time it is drunk. Too bad if the consumer prefers it that way.

It's only American scientists who still believe that this kind of wine is good for the heart.

Name Your Poison

All that remains is to sell the concoction on the pre-arranged date and wash the grape baskets, presses and vats in preparation for the next harvest. Every year, some 6 million cubic metres of water are used for washing and then poured away (with 6 per cent grape juice content) into drains and rivers: the treatment plants, where it ends up, simply can't cope. Even under normal circumstances, they only recycle 60 per cent pure water, which means that fish are unable to survive. Because the water is too rich in waste matter, there is a proliferation of micro-organisms to break down the grape juice, the biochemical reaction uses up a large amount of oxygen and the fish are asphyxiated.

Fortunately, producers of more than 50,000 litres of grape juice per year are required by law to treat their own waste water by installing a purification plant or sometimes simply by spraying it over the fields (which is as good a fertiliser as any). That's the least they could do when you think that a single co-operative generates as much pollution as a town of 80,000 inhabitants.

Making it Go Further

Some wine makers are more like Sorcerers' Apprentices than true practitioners of the noble art of viticulture. Whole regions of France have been spoiled as a result of their production methods.

If the wine is in danger of being too bland, you can use a certain type of yeast which will give it a banana flavour (flavourings are not permitted). A short time in young wooden casks will 'round out' the taste.

Perhaps it isn't sweet enough? Then add some sugar (unless it has previously been fortified with tartaric acid). Preferably grape sugar, known as 'rectified concentrated must': there are several French and Italian vineyards which produce only that.

If it lacks acidity, fortify it with tartaric acid (unless it has already been chaptalized). If it lacks body, a bit of glycerol will put it right. There are also more sophisticated techniques: for instance, preheating, vacuum treatment and de-pressurisation. An all-in-one method of improving the body, bouquet, structure and colour of the wine (lesson No. 42 for the wine maker of the future).

The 'mutant vine' is here.

And if you want to be able to transport it safely to the ends of the earth (to Tokyo, for example), you'd better pasteurise it before it gets shaken about. Over there you can still find a few suckers who'll pay for £30 or £40 a bottle for some revolting brew which will give them a herruva headache...

Then remember to say on the label that the wine (if you dare call it that) should be drunk cool, and don't forget to number the bottles (that looks good), think of a name like *'Sainte-Émilie-Hons'* (which will deceive anyone whose spelling is as poor as their knowledge of wine) and stick some kind of cachet on it, like 'authorised vintage' or 'aged in oak casks' or 'Gold Medal vintner's personal reserve' (it won't cost you anything).

When you've done all that, make sure the bottles are stood upright on the most brightly lit shelves (the antithesis of good storage procedure) and you know you've produced a wine that really deserves the 'appellation' dishwater.

Scraping the Barrel

'What a load of rubbish! That sort of thing only happens abroad,' the industry professionals will say. 'French wines are still the best in the world. You only have to taste an *AOC (appellation d'origine contrôlée)* wine to appreciate that!'

AOC wines account for half of all French wine production. Introduced in the 1930s, the classification is awarded by the *INAO (Institut national des appellations d'origine)* and in principle (as well as guaranteeing the place of origin and imposing a production quota) certifies the grape varieties and wine-making methods used and assures a certain level of quality. But, over the years, couldn't a degree of leniency have crept in and allowed a few wines to be awarded an *AOC* which didn't really deserve it, or more probably one or two to retain their *AOC* which no longer deserved it?

It is hypocritical to go on maintaining that nobody outside the borders of France knows how to make good wine. The actual situation is now so confused that no one knows where they stand: wine specialists doing blind tastings of wines from France and from other countries can't tell them apart!

On top of all that, there are the local vinos with false labels which so often make a name for themselves. Only recently in Languedoc, the *DGCCRF* (remember them?) uncovered a huge trade in fake *Châteauneuf-du-pape*. It had been produced by discolouring ordinary red and was virtually undrinkable. They managed to make several million francs out of it all the same!

In 1994, of the 3,733 wine-making establishments inspected by the Fraud Squad, '1,380 were issued with warnings and 259 were charged with falsification, misleading publicity, unauthorised blending, deception, improper use of appellations, misstatement of crop sizes and illegal irrigation.'

The figures speak for themselves.

Labels that Won't Wash

Although it doesn't say much, the label on a proper wine-bottle is more revealing than it might at first seem. All the wording which is required by law appears at the bottom. Such as, in the case of French wines:

> – The name of the specified region followed by one of the following four categories: *appellation d'origine contrôlée (AOC)*,

vin de qualité supérieure (VDQS), vin de pays or *vin de table*.

– The name and address of the bottler. *Mis en bouteille à la propriété, au château* or *au domaine* indicates that this is done by the grower himself. Anything else means that it is done by a wholesaler.

– The volume in centilitres and the alcoholic content.

All other information is optional (and not always fanciful), whether it's the name of the wine, its year or any sort of classification it might have.

As for the picture which usually adorns the label, the law says it must not 'create confusion in the mind of the purchaser as to the nature, origin or quality of the product'. A subjective judgement if ever there was one. The words grand vin de... are in no way a guarantee of quality, any more than descriptions such as 'from old vines', 'aged in oak casks', 'first vintage', 'reserve vintage', 'traditional vintage' or any other kind of vintage. Any classification a wine might have is specific to the region it comes from.

Beyond that, it is certainly a pity that wines, unlike food products, do not have to have all their ingredients listed on the label. But even those wouldn't tell you that wine growers in the Champagne region were supplied by a certain town with special industrial 'sludge', ostensibly 'to enrich and stabilise the soil'. There are no details as to what concentrations of dangerous metals might end up in the resulting 'bubbly'.

Don't count on the dealers to tell you either: they have far too much at stake to commit any such indiscretion.

Your Health!

What is true of wine is equally true of spirits. But, of all alcoholic drinks, it is undoubtedly beer which makes most use of the latest food industry technology – i.e. of the most additives of all kinds.

Bromelin or papain to clarify it or to protect it from cold, ascorbic acid as an anti-oxidant, propylene glycol alginate (up to 0.1g per litre) to control frothing, all sorts of miscellaneous colourings, etc, but not a single word about any of them on the can.

Whatever French breweries might say, the only beer made exclusively from malted barley, hops, water and yeast is German. It is a tradition which dates back to 1516, when a so-called 'purity law' was decreed, which is still in force today.

Sadly, if the French want to taste it, they actually have to go to Germany (the export 'brew' is hardly any more genuine than its competitors). A taste of their own medicine, in a sense, since the French used the excuse of free trade in the European Community to justify their right to sell their own beers in Germany. Without having to abide by any purity law.

With or Without Gas?

When it comes to those fizzy drinks which are so popular nowadays, once again, anything goes – or practically anything. They are usually made by diluting between one and three grams of 'soda ash' per litre of carbonated water: the ash consisting of soluble and emulsifiable flavourings, colourings and preservatives (including antioxidants and gum arabic). To cap it all, the formula varies according to how the drink is to be packaged and sold.

Fruit juices aren't much better: water and sugar can be added to make them less bitter or acidic. That leaves us with water. But tap water can have abnormally high levels of nitrates in certain areas or an excessive dose of lead from pipes. Even bottled water can be affected by accidental chemical or micro-biological pollution. The *DGCCRF* are keeping a close eye on it and have closed down more than one source.

14

MIND THE COLD!

From cooling machine to refrigerator via cold stores and deep freezes, we soon reach the lowest degree of food preservation.

Refrigerated food is kept at between 0° and 4°C, while frozen food is gradually cooled to at least -18°C and deep-frozen food to -38–40°C. This doesn't mean that any food can be kept at any temperature. From producer to consumer there are certain basic and common-sense rules which must be observed. Or at least ought to be.

Managing to Keep Cool

Fridges don't work miracles: mince will keep for two to three hours, raw fish for a day (provided it has been gutted and rinsed), poultry and other kinds of meat for two to three days and vegetables hardly any longer. Top whack!

What's more, the food must be absolutely fresh to start with. But readings taken of refrigeration units in shops (large and small) sometimes reveal temperatures of between 11° and 16° instead of 4°C. Fraud Squad reports show that 'the most frequent irregularities [are found in] refrigerated products'.

Even when the fridges are working properly, one should always resist the temptation to take a packet from the top: these are the ones which are most exposed to temperature variations. Not only have they probably already been handled several times, but they're also there for a reason. The shopkeeper puts them within easy reach (fair enough) in order to get rid of them first, because they're getting close to their sell-by date.

If you must buy perishable foods packed by the retailer, you'd do better to avoid the less busy shops where the shelves aren't restocked often enough. That goes for raw mince, grated carrot, mixed salads, offal, 'home made' mayonnaise or chocolate mousse, poultry, shellfish and seafood. Etc.

Unless, of course, you like to live dangerously: bacteria and fungi remain active down to -10°C!

Once you take the food out of the fridge, it starts to warm up and in some cases deteriorates very quickly. When it reaches 5°C, salmonella bacteria start to multiply, and at room temperature they double in number every twenty minutes. A ham wrapped in cellophane will take only half and hour to reach 17° or 18°C.

What about the final stage in the process –

your own fridge: can you honestly say that it is regularly cleaned and precisely adjusted? There's only one way to check: using a thermometer. At the top, the middle and the bottom and in each compartment, because the most perishable foods should logically be kept in the coldest part.

Imagine if the *DGCCRF* were to check household fridges as well!

Frozen Stiff

What's true of cold food is all the more true of frozen and deep-frozen products.

Food cannot be sterilised by freezing or deep-freezing: at best, very low temperatures can only postpone the development of bacteria and pathogenic enzymes. If the cooling process is interrupted at any stage, even partially or for quite a short period, the consequences are therefore irreversible.

For example, a packet of frozen beans contains roughly a thousand bacteria per gram. By the time it has thawed out for twenty-four hours at room temperature, there will be 40 million bacteria per gram. So a 100g portion will contain, yes, 4 billion bacteria. Is that all?

The French army once had a bad experience with a lot of mince which hadn't been properly frozen and as a result was contaminated: the Supply Corps discovered that it contained tens of millions of bacteria per gram. History doesn't record whether these lousy victuals were decommissioned before or after being tested in action. Military secret.

Even though the links in the chain appear to be strengthened a little each year, there are always problems with keeping things cold. Out of more than 4,000 checks in 1994, the Fraud Squad found some 320 cases of failure to meet standards. And out of a total of approximately 100,000 micro-biological tests, they discovered at least 20,000 infringements of hygiene or temperature regulations (only 30 per cent of which had been sanctioned)[1].

The worst thing that can happen is for food which has started to defrost to be refrozen. This can damage its structure, if the formation of ice doesn't actually make it explode. Pockets of liquid can also form inside the food where enzyme decay and oxidisation can occur.

If you don't want to add to the risks, you'll have to make sure you that when you buy deep-frozen food you check the temperature of the freezer (it should have a thermometer in it), carry it in an insulated bag and get it home as quickly as possible.

Mind the cold!

---

[1] Admittedly, the *DGCCRF* did specify that these statistics should be interpreted with caution. As we have noted, the officers of this noble institution have a knack of inspecting the most 'sensitive' locations first (if only to save time), which means that the figures we have quoted might be slightly distorted.

Going to Work on Some Eggs

Here's another case of papering over the cracks. With fresh produce it's always best to observe the sell-by date printed on the packaging, but that doesn't mean you should put too much faith in those 'freshness tags' which seemed at the time like a step in the right direction.

The magazine *Que Choisir* set the record straight when, in June 1995, it exposed them as just another marketing ploy. These tiny lozenges, which appeared on food packaging, were supposed (or so we were told) to indicate how fresh the contents were by changing colour, but in fact guaranteed nothing at all. The magazine tested five samples of ham, five of cheese and the same number of 'Class 4'[1] salads, and found that whether they had been kept at the proper temperature of 4°C or less, or left out at 25°C for five hours, 'not one of the tags worked properly': they only just began to change colour long after the food had gone to the dogs.

While on the subject of food poisoning, it wasn't so long ago that eggs headed the hit list. But a new ruling has provided greater safeguards: eggs may only be sold as 'freshly laid' during the week after packing. After that, they must go into the 'fresh' category, provided that they can be con-

---

[1] This classification relates to the method of production rather than to any qualitative criteria. 'Class 3' products include all deep-frozen food and 'Class 5' comprises catering packs, of both raw and cooked food, for refrigeration only.

sumed within a further week. What the label doesn't necessarily tell you is what happened before the eggs were packed. For example, how the hens that laid them were bred, reared and fed.

But old habits die hard, and unfortunately a legal document isn't a magic wand. The *DGCCRF*'s annual report for the year 1995 makes instructive reading. We quote the following excerpt from it: 'The findings of the 1993 inquiry into eggs and egg products, namely that the samples tested were of inferior quality and a hazard to health, prompted a further series of tests in 1994. Inspections carried out at twelve manufacturing plants, twenty-two battery farms and twenty-one hatcheries or small breeding establishments showed up persistent fraudulent practices. Proceedings were instituted against six manufacturers of egg products, three of whom were arrested for reprocessing dried egg whites and one for using fertilised eggs, which is illegal. Four producers were prosecuted for selling eggs which were unfit for consumption.'

Let's take another look at those figures and put this 'inferior quality and hazard to health' into perspective: out of fifty-five producers inspected, ten come a cropper, which means 18 per cent of them are on the fiddle.

Pull the Udder One!

In comparison with eggs, French milk, whether

pasteurised (heated to between 72° and 85°C), sterilised (115° - 118°C) or UHT (150°C), begins to look like a model of good behaviour. Nevertheless, it must be admitted that the terrible poisoning it can cause, which can hardly be blamed on the milker (the cow, not the maid), is no longer acceptable.

Even more worrying perhaps are some of the technological 'advances' which have been proposed. If it's no longer a case of genetic manipulation as such, but of 'biotechnology', it's hardly any more reassuring, nor does it prevent certain highly specialised and well-informed publications from putting forward clever ideas which could prove extremely profitable for the producers concerned. Time will tell.

The biotechnology in question (which has been carried out exclusively in the United States) has involved identifying in cattle and isolating 'the gene that codes for the synthesis of the hormone which stimulates lactation. The gene has been transferred to a bacterium which is thereby enabled to secrete the hormone. The bacteria are reproduced in a fermenting apparatus, purified and made available commercially by X at a competitive price.'

Cows inoculated with these bacteria will gush milk like geysers.

The same article continues: 'At the end of the day, it will be impossible to distinguish this type of milk from the more traditional product. Even the labelling will be the same, due to a legal loophole on the other side of the Atlantic which still allows this to happen. Consequently, there has been a significant movement among consumers to boycott milk and milk products altogether until there has been some ruling , if not on the products themselves, at least on the question of labelling.'

Make no mistake, it won't be long before the laws of free competition laid down by the GATT and its successors will come home to roost. Nor will there be any hesitation about trying to force them on the European market by presenting us with a fait accompli, which governments will ratify retrospectively while the European Commission smoothes out the differences between national regulations from the bottom up on the pretext of 'harmonisation'.

To whose advantage? Work it out. If every dairy farmer is producing 20 per cent more milk, who'll get the cream? What it amounts to is that the only thing still holding Europe back from following resolutely in America's footsteps is fear of a consumer rebellion.

It's what you might call a burning issue in the world of cold food.

Far from cold (or fresh for that matter) is powdered milk, which dissolves instantly thanks to a dose of lecithin and which is used in Eastern Europe (where neither hormone treatment nor the use of antibiotics is subject to any sort of control) in the manufacture of cheap yoghurt.

Plumbed Tomatoes

Could it be, as someone once said of the ravages of time, that being 'well preserved means having nothing left that's worth preserving'? Now there's a question. Especially when you discover, for example, that more than half of the tinned (or dried) mushrooms on sale are unfit for consumption. So sayeth the Fraud Squad. Or again when you read a headline like this: 'CANS CONTAMINATED – 213,000 PEELED TOMATO TINS SEIZED.'

It was the *DGCCRF* who found these particular tins in a depot at Bobigny, but it was more like Davy Jones' locker than Aladdin's Cave, since they were all 'badly rusted'. As it happens, 450,000 similar containers had been bought from a Vaucluse producer who had been a victim of the notorious 1992 floods. Whoever decided to save himself from drowning by re-labelling 100,000 of the damaged tins and leaking them onto the market, must have been able to walk on water.

450,000 tins minus 213,000 impounded and 100,000 got rid of equals 137,000 unaccounted for. According to the *DGCCRF* itself, they were still being sold at the end of 1994.

15

MIND THE HEAT!

Blinded by science, we end up in the soup.

France has more restaurants per capita than anywhere else in Europe, with eighteen for every 10,000 inhabitants. Which is hardly surprising when you think how much the French love their food and how famous their cooking is.

In fact, its reputation is as high as the sums of money spent on it. For each restaurant meal, 50 per cent of consumers pay (unless they're being paid for) up to £6, 30 per cent spend between £6 and £12 and the remainder share the more extravagant bills. And in total, every year, the French eat out some 5 billion times. It's enough to make you feel quite ill.

Of course, in more than 60 per cent of cases, we don't mean a five-course meal in a three-star *Guide Michelin* restaurant. Nor even a cosy candle-lit dinner for two. We're talking about mass catering – in other words the ransom that must be paid to a system which obliges people to work further and further away from home.

There's no doubt that such a fantastic market excites the imagination of food manufacturers, restaurant chains which repeat the same formula ad nauseam (they already have a 20 per cent slice of the cake), fast-food outlets which are destroying the traditional café, and industrial caterers who supply to factories where workers line up with their trays at the self-service counter like prisoners. Not forgetting school canteens, of course. But their imagination always tends in the same direction: towards lower production costs and standardised taste. A sort of inverted creativity.

Just as artificial colouring can stimulate the appetite, so all that distinguishes one restaurant from another when it comes to paying the bill are the setting, the ambience and the service. In the back kitchen, the so-called 'chef' operates more and more on auto-pilot.

But is it Art?

It's all in the decor. The outside is either bistro-style or Mexican gaucho-style, Rococo or Hollywood studio-style to attract the customer. Or should we say the fall guy, since on the inside, where the ovens and the fridges are, there's no time for fancy cooking any more.

Because time is money.

MIND THE HEAT!

Still, they have to give you something to eat, which usually means something pre-prepared, pre-packed and pre-cooked, not to say pre-digested. The era of 'cooking by numbers' has arrived. It's more like assembling than cooking, because all you need to know is how to work a tin-opener or a pair of scissors and how to press a button.

A rep called in a few days or weeks ago. You naturally took the opportunity to place an order from the usual catalogue by ticking the numbers next to the pictures. Then, just this morning, a delivery van arrived with the goods (more likely in cardboard boxes than on carcasses).

Put them away and start getting the place ready. There's just time to make sure the tables are properly laid before the first customers arrive.

'What do you recommend today?' asks a regular, 'I'm in a bit of a hurry...'

'In that case, may I suggest the traditional local pot-luck bourgiglobule? It comes with country vegetables in a farmyard sauce. I'm sure you'll like it...'

It's on for the dish of the day with its thingamabob vegetables. The kitchen immediately springs into action:

'A farmyard globule for number three!' yells the waiter: 'And I want it yesterday!' he adds (if he's in a good mood).

The cook goes into a flat spin. Farmyard globules don't grow on trees.

Stage one: open the tin of powdered farmyard sauce and mix it with water (not too much and make sure there's someone to stir it), then pour it into a saucepan and heat it up.

Stage two: drop the plastic bag containing the individual portion of vacuum-cooked bourgiglobule into boiling water and find those blasted scissors to cut it open and pour it out in one single action.

Stage three: check that the country vegetables have thawed out properly and re-heat them in the microwave.

Three easy stages, and all that remains is to 'assemble' the globule, the vegetables and the sauce like building blocks on the plate and sprinkle it with parsley (no need for it to be fresh, let's be honest), which is what's known as 'putting the icing on the cake'.

It's all done in three minutes flat. Bravo, a work of art! The customer will be none the wiser and might even leave a tip (in return for the one he has just eaten), provided you've wished him *'Bon appétit!'* with the proprietary smile.

In the Soup

Powdered omelette mixes, sauces and puddings; dehydrated soups and sauces, dried prawns and granulated gravy; frozen pastry, chips, venison, pheasant, salmon, coquilles Saint-Jacques, mushrooms and herbs; reconstituted meat juice; vacuum packed beans, meats, fish and pizzas; tinned lasagne, wrapped vegetables and packet salads: anything goes in the world of 'ready-

MIND THE HEAT!

when-you-are' products, peppered with additives and 'technological assistants', and all the cook has to do is to keep them cold or heat them up.

The majority of restaurant chains follow the same ritual. But it isn't such a new idea: the catering business has been pouring out mass-produced sauces for a long time now. The kitchen is no longer where you think it is: instead of being behind the restaurant, it is located 'centrally', often hundreds of kilometres away. That doesn't mean it isn't still 'integral', since it is part of the same group as the restaurants it serves.

Take this typical example: fish are bought at auction from a tiny port in Brittany, preferably when prices are at their lowest, and put in a cold store. Then they are all 'processed', cooked, packed and trucked off to the restaurant. Anyone who doesn't like fish can have grilled meat instead. Well, 'grilled' is perhaps a slight exaggeration. Grilling takes a long time and the restaurant doesn't actually have a grill, only microwave ovens, so the surface of the meat is 'treated' with caramel. It looks the same and it's a lot quicker.

What's the advantage? Considerable. Enormous, in fact, in business terms. In the kitchen, it limits waste in defrosting to order, at the same time as economising on staff and saving time on preparation, cooking, re-stocking and washing-up. In the restaurant itself, where on the other hand more money is spent than before, it means being able to provide a varied menu and offer 'seasonal' produce all the year round. Plus a guarantee of consistent quality, even allowing for the chef's occasional moods.

The net result is that the cost of the food is in inverse proportion to the prices on the menu.

And perhaps we should say a word about the menu. Just as it tells you nothing about the true nature of the food on offer (or the chemical ingredients it contains), so it conceals from you whether such and such a dish has been pre-cooked, pre-heated, vacuum-packed or deep-frozen. In fact, the only things it can't say are 'home-made', '...of the day' and 'chef's special': that would be misinformation.

In other words, silence is golden. In this case, honesty is definitely not the best policy.

'Shut Up and Eat Up!'

Not all of France's 100,000 restaurants treat their customers in the same way, but we shouldn't delude ourselves: although there are no official figures, more and more of them are being tempted to take short cuts.

Needless to say, refectories and canteens are hardly the most scrupulous of places either. School dinners alone account for a billion meals a year, which makes them a £2.5 billion business. The meals that are served are 'price-censored', as it was so nicely put by the boss of one of the dozen companies competing for the 'processed food' market. Indeed, price ceilings are sometimes so low that only a third of schools, colleges and universi-

MIND THE HEAT!

ties observe the official nutritional recommendations: that they should supply at least 40 per cent of the daily requirements of children and young people.

There's more choice in the business sector, where hundreds of lorries deliver 'made-to-measure' restaurants day in day out. Eight or ten starters, two or three main courses, an optional 'grill', eight cheeses and ten desserts is the usual minimum for a large office or factory canteen. Or a hospital. Even the canteen at the European Parliament in Strasbourg works that way, although it has the biggest catering budget and just about the best 'assembly kitchen' food of any establishment in France.

The culinary arts don't benefit much from all this, but it's said to be an improvement on the sensory GBH perpetrated by the old roadside restaurants. The EC standard ISO (International Safety Organisation) regulation 9002 imposes Draconian hygiene regulations, quality and safety checks are becoming more frequent and outbreaks of food poisoning increasingly rare. The real health problems will come later, when the additives start to take effect.

16

A QUICK DIG IN THE RIBS TO HELP IT ALL DOWN

*A long life
and a healthy one?*

Our life expectancy is constantly increasing, we're told. And this is the crucial argument every foodstuffs manufacturer uses to persuade us that the planet is not as badly polluted as we think and that modern food is becoming more and more healthy, nutritious, well-balanced and appropriate to our needs.

It's rather a simplistic argument, but why not take it at face value and regard it as a good omen?

Possibly because it's ominous. As a leading demographer has stated, your life expectancy is obviously no guarantee that you will actually live that long, and it is affected by many factors other than nutrition. So it doesn't need a prophet of doom to question the validity of the argument. What we know is that within the space of two generations standards of living have improved considerably: greater comfort, more leisure time, improved hygiene, fewer industrial accidents and an easier way of life. People don't have to fetch water from a well any more, houses are better heated and childhood ailments are more easily cured. Infant mortality has fallen dramatically, and medicine has made enormous progress.

And yet more and more people are dying of cancer, even though they're taking longer and longer to die.

The relationship between food and demography is further complicated by several other factors. For example, the contemporary oestrogen plague is particularly revealing. In this case the stabilisers and anti-oxidants used in plastics, like nonylphenol are suspected of generating oestrogen. They could be the cause of problems with babies whose testicles won't drop and of a growing number of cases of testicular cancer in adults (in Denmark, there has been a 300 per cent increase in the last ten years!). They might also be responsible for the degeneration of sperm, a phenomenon which is becoming extremely widespread. Apparently, 75 per cent of spermatozoa in the European Community have either two tails or two heads or swim round in circles. That's if they aren't up to some monkey business of their own.

But nonylphenol is in everything. It's used in the manufacture of paint, detergent, shaving foam, shampoo and soap, but also in those famous pesticides. At a concentration of 30 micrograms per litre, it increases the activity of female hormones a thousand times. And since it is not unusual to find concentrations of up to 50 micrograms per litre, we can safely predict that by the year 2010 Woman shall have inherited the Earth.

It is already happening among the fish populations of some North American lakes and rivers.

That leaves the foodstuffs manufacturers with only one justification for their production methods: overpopulation. 'How else are we to feed an ever increasing number of people?' they will ask, hand on heart. And anyone with other scruples must either be living in the past or have a chip on his shoulder (or else a full stomach).

Another specious argument, because how can you justify the race for profit in terms of the population explosion, when the food industry is already theoretically able to supply every member of the human race with the required 2,500 calories a day? And why doesn't it? For what inadmissible economic reasons does it leave more than a billion men, women and children dying of malnutrition and hunger?

If only the plunder of the seas, the frenzied exploitation of the land, the massacre of nature and the tremendous threat to health we all face were for the sake of preventing people starving to death before our very eyes, as they do every night on the News at Ten, in between adverts for supposedly slimming foods and calorie-controlled diets! It would be a fairly short-sighted policy, but at least the planet's limited resources wouldn't be going to waste.

Since this is not the case, it has to be said that the food industry is anything but altruistic. Its fine words are an inadequate disguise for more mundane preoccupations like short-term gain on investment.

In the long term, it leaves future generations to sort out the mess as best they can.

But whose fault is it? The producers' or ours as consumers? It may be the eternal conundrum of the chicken and the egg, but why should we forever give in to the fatal temptation to buy the average, cheap product, thereby encouraging manufacturers to make them even more average and cheap?

What is stopping us from changing our lifestyle and our eating habits? How can we be so short-sighted as to accept that the total number of plants selected and cultivated for mass food production is now as low as 150, of which merely thirty provide nine-tenths of our vegetable intake? What are we going to leave to our children? *'Après nous le déluge'*?

Lots of questions, the answers to which all point to our own complicity in what we witness as we inspect the contents of our plates and glasses.

If, instead of picking a perfectly smooth, yellow apple which is infested with chemicals, we were to choose a smaller, less sweet and slightly deformed apple which is so much healthier and safer; if we were to buy only naturally greyish pork rather than artificially pink pork, we might

A QUICK DIG IN THE RIBS...

yet be able to stop this crazy juggernaut.

At least it would be worth a try. Remember what happened to hormone-treated veal. And remember also that when natural colourless mint and grenadine juice appeared on the market, it was because of our own stupidity that it didn't sell.

If there's something rotten about the food world, it's up to us to put it right. It's never too late, even though the die may already be cast. To save what can and should be saved and to halt this continued suicidal waste, we must first change our own behaviour.

We mustn't be so complacent. When we're eating out, we should bang on the table and demand to know exactly what is being put on our plate. When we're out shopping, we should give ourselves time to read the labels. And if manufacturers are not obliged to list all the additives in ingredients which make up less than 25 per cent of a particular product (for instance, the supposed sausage in a supposed meat pie), then at least we should take the trouble to compare different brands.

Even the order in which the ingredients are listed is significant, because they are arranged according to quantity: the largest at the top. And, as far as possible, we should choose the least 'doctored' products.

Such efforts may seem like a drop in the ocean, but at the very least our own health will benefit. And if we don't care about ourselves, let's think of our nearest and dearest who are eating the food we buy them: if nothing else, let's stop poisoning our children!

Let's not be taken in by products on the shelves and in the freezers which are made to look attractive in order to conceal their true identity. On the contrary, let's give priority to small producers' goods such as we see on market stalls where a few vegetables and some fruit compete for attention: the quantities they grow don't justify the use of expensive pesticides.

Let's encourage and reward the efforts of those who insist on growing 'organically'; let's grow our own vegetables if we can, or buy them straight from the farm. Above all, let's do our own 'processing': in other words, cook the food ourselves instead of letting the manufacturers do it for us.

Then let's spread the message to our friends and relatives. It isn't a question of sectarianism or any kind of militancy, but simply of making people aware of what is at stake. That way we can get organised and, if necessary, establish our own supply lines.

Just as one plus one equals two, so our efforts will be combined and multiplied by others until they do indeed carry some weight in a market which is constantly studied, analysed, dissected and calculated to the nearest tenth or even hundredth of a per cent by the food suppliers and their various support groups.

When their sales and profits start to decline, they'll want to know why, and in the end they'll have to do something to regain their share of the market. Wherever we do our shopping, let's do likewise. Let's demand explanations, let's ask embarrassing questions and not be content with vague answers. After all, it's in the retailers' interest to keep us happy.

If each member of every family does the same thing, at the same counters, in front of the same shop-assistants, the message might eventually arouse some interest and be passed on up to the management. Once they hear the alarm, they might even be more amenable to our demands!

In short, we must no longer allow ourselves to be dictated to by others. We must stop acting the way they want us to, like a herd of sheep on its way to the slaughterhouse. We must listen to and follow the advice of consumers' associations. We must buy their publications and read their reports. That will give them all the more power to pursue their enquiries, expose scandals, enforce regulations and put pressure on those who make them to improve them.

As for politicians, we must harass them, interrogate them, pursue them, hound them into a corner. Make them stop giving in to increasingly persistent, aggressive and greedy lobbies. Make them revise the electoral laws, which are responsible for the over-representation of the most powerful agricultural interests. Make them harmonise regulations from the top down by establishing international (or at least European) vice squads. And since alternative solutions do exist, make them support these as a matter of urgency.

American, British and French scientists are already pretty well engaged in a frantic 'clean-up' campaign. Everywhere they're breeding waste-eating bacteria, aphid-eating ladybirds and single-celled organisms which will devour plastic – all of them environmental developments aimed at improving our living conditions. Let's make sure that these experiments find their way from the laboratory to the land.

Once every single gene has been identified and the genome project has been completed, it may be possible to blow the whistle on polluting industries and to increase yields without recourse to an arsenal of chemicals. As long as these discoveries don't fall into the hands of people who can afford to buy up the patents and hide them away in a bottom drawer.

And since 1996 marked the 50th anniversary of the French National Federation of Agricultural Enterprise Syndicates (the *FNSEA*), let us take its leaders at their word when they talk of entering into 'a contract with the nation concerning the position and the role of agriculture'. Because that is exactly what we need, now that it is dangerous to pull up a carrot in a field, suicidal to pick an apple off a tree, an act of heroism to bite into a steak; now that cows no longer recognise their own milk and bakers no longer know what their flour is made of (if bakeries aren't already little more than ovens for cooking ready-made lumps of dough).

It's high time we put an end to the incredible wastefulness, the unbearable laziness, the monstrous blindness which has bound and gagged us and left us hostages to the money-makers, to a system which is heading for disaster and will surely take us with it. Otherwise, our only consolation will be that, having unwittingly consumed so many preservatives which continue to be effective long after their sell-by date, our corpses will take at least twice as long to decompose as they would otherwise.

TO FINISH

ADDING UP THE BILL

SOME RECOMMENDATIONS

DON'T FORGET THE TIP!

ADDING UP THE BILL

A short list to guide you through the maze of E numbers and extricate you from multiplying additives.

This is a selective list which does not claim to be a chemical dictionary of food additives. It is merely intended to illustrate the composition of some of the substances with which we are inundated, and has been put together with the help of researchers at the Institute of Physics and Chemistry in Paris (who themselves had to refer to specialist encyclopædias in order to find their way through the labyrinth of European codification).

Permitted Colours

E100: Curcumin (yellow).
E101: Lactoflavin or Riboflavin (yellow).
E102: Tartrazine (ADI of 7.5mg per kilo). Used for colouring cheese rinds, cold meat skins, ice-cream, sweets, cakes, etc.
E104: Quinolene Yellow.
E110: Sunset Yellow.
E120: Carminic Acid or Carmine of Cochineal. Made from South American insects, it is often used as a glazing for salted cooked meats.

E122:	Azorubine (red).
E123:	Amaranth (red). In France used only for roe (caviar) and fish substitutes.
E124:	Ponceau 4R (red).
E127:	Erythrosine. Red colouring for sausages, non-alcoholic drinks, ice-creams and sorbets, cakes, chewing-gum, sweets, etc.
E131:	Patent Blue V.
E132:	Indigotine or Indigo Carmine.
E140:	Chlorophyll (green).
E141:	Copper complexes of chlorophyll and chlorophyllins (olive-green).
E142:	Acid Brilliant Green BS.
E150:	Caramel (brown).
E151:	Brilliant Black BN.
E153:	Carbon Black or Vegetable Carbon.
E160:	Carotenoic acid, Bixin or Carotene (yellow/orange). Often used as a colouring for pasta.
E161:	Xanthophylls. A group of yellow pigments obtained, for example, by thermic coagulation of proteins in crushed alfalfa juice, which will colour eggs, chicken, biscuits, sweets and cakes, etc.
E162:	Beetroot Red or Betanin.
E163:	Anthocyanins (red, blue or violet).
E170:	Calcium carbonate (white).
E171:	Titanium dioxide (white).
E172:	Iron oxides and hydroxides (yellow, red, orange, brown and black).
E173:	Aluminium.

E174:	Silver.
E175:	Gold.
E180:	Pigment Rubine. Used solely for colouring cheese rind

Preservatives

E200:	Sorbic acid. A sugar derivative often used in sauces, for example.
E201-203:	Sodium, potassium and calcium sorbates. Used to obtain permanent solutions. E202, with added citric, lactic, tartaric or acetic acid, is also used as an anti-fungal agent on sausage skin, for example.
E210:	Benzoic acid. A petroleum derivative included in some fruit-flavoured fizzy drinks.
E211:	Sodium benzoate. Used in conjunction with E202 in meat, fish or vegetable salads with emulsified dressings.
E212 & 213:	Potassium and calcium benzoates.
E220:	Sulphur dioxide.
E221:	Sodium sulphite.
E222:	Acid sodium sulphite.
E223 & 224:	Sodium and potassium metabisulphites.
E226:	Calcium sulphite.
E227:	Calcium bisulphite. A common preservative in wines, ciders, jams, concentrated fruit juices, dried fish products, etc.

E249 & 250: Potassium and sodium nitrites (maximum permitted quantity: 0.2 per cent).

E251 & 252: Sodium and potassium nitrites (saltpetre). The ADI is 3.65mg per kilo, but a court ruling has permitted a concentration of 50mg per litre of drinking water. Potassium nitrite is commonly used in cold meat skins.

E260: Acetic acid (derived from ethyl alcohol).

E261: Potassium acetate.

E262: Sodium diacetate.

E263: Calcium acetate.

E270: Lactic acid (derived from lactose).

E280: Propionic acid. A superior form of acetic acid.

E281-283: Sodium, calcium and potassium propionates.

E290: Carbon dioxide (derived from carbonic acid).

E296: Malic acid. A flavouring extracted from fruit.

Anti-Oxidants

E300: Ascorbic acid or Vitamin C. Inhibits browning in cut fruits.

E301 & 302: Sodium and calcium ascorbates.

E304: Ascorbyl palmitate.

E306: Extracts of natural origin rich in tocopherols. Extracted from wheat

ADDING UP THE BILL

	germ, tocopherols are unstable elements used in medicines to combat the effects of the menopause and sterility. (There are a lot of them in corn-flakes.)
E307-309:	Synthetic alpha-tocopherol, gamma-tocopherol and delta-tocopherol.
E310-312:	Propyl, octyl and dodecyl gallates.
E320 & 321:	Butylated hydroxyanisole (BHA) & Butylated hydroxytoluene(BHT). These synthetic anti-oxidants are derived from petroleum and also used as preservatives (in chewing-gum and dehydrated potato flakes). Known to be toxic, they have recently been identified as carcinogens.
E322:	Lecithins (also used as emulsifiers, stabilisers, thickeners and gelling agents). Extracted from soya beans, these are natural fat concentrates (containing at least 56 per cent of phospholipids). Lecithin is the only emulsifier permitted in the manufacture of baguettes.
E325-327:	Sodium, potassium and calcium lactates.
E330:	Citric acid. Originally isolated from eggs, this acid is also found in citrus fruits and soya beans. It is often used to preserve the whiteness of asparagus, salsify and palm hearts, and as an emulsifier (a fatty substance which adds volume).

E331-333: Sodium, potassium and calcium citrates.
E334: Tartaric acid. Used as an acidity corrector, it is also a laxative and flour stabiliser (e.g. in photographic film).
E335 & 336: Sodium and potassium tartrates.
E337: Sodium potassium tartrate.
E338: Orthophosphoric acid. A chemical composition artificially produced by decomposing certain fruits and used as a paint stripper and fertiliser as well as a preservative in the world's most famous fizzy drink. (Causes corrosion in large doses.)
E339-341: Sodium, potassium and calcium orthophosphates. Anti-oxidants as well as emulsifiers, stabilisers, thickeners and gelling agents.

Texturizing Agents
(emulsifiers, stabilisers, thickeners and gelling agents)

E400: Alginic acid. Extracted from seaweed, this emulsifier is much used in certain beers (maximum permitted dose 1.5 per cent).
E401-405: Sodium, potassium, ammonium, calcium and propylene glycol alginates.
E406: Agar-agar or artificial gelatine. Extracted from seaweed and used in cold meat glazing, photographic products and cosmetics.
E407: Carrageenans (Irish Moss). Natural

extracts of seaweed 'not subject to particular dosage restrictions other than that of the quantum sufficit, the amount required to achieve the desired effect'. Carrageenans will retain water with the aid of gelatine (but also concentrate heavy metals). These additives are injected into lower quality hams (maximum permitted dose 0.5 per cent) and used in cooked sauces, ice-cream, milk puddings, etc.

E410: Carob bean gum. Often used as a gelling or bulking agent, for example in baking flours. Its maximum permitted dose, which has recently been lowered, is 0.5 per cent.

E412: Guar gum or flour, used as a thickener, for example in chestnut purée.

E413: Gum tragacanth or dragon.

E414: Gum arabic.

E415: Xanthan gum. A microbial gum used either as a gelling agent or as a stabiliser, for example in mustard (other than Dijon mustard). It has a maximum permitted dose of 0.1 per cent of net weight.

E420-421: Sorbitol and mannitol (Manna sugar). Artificial sweetener used for example in certain types of 'diet' chewing-gum.

E422: Glycerol (acid extracted from oils and fats). Very commonly used in

	soaps, solvents, diluents, anti-gelling agents and solvents, it is also one of the basic ingredients of nitroglycerine and acts as a 'stabiliser' in wine.
E432-436:	Polyoxyethylene sorbitan monolaurate, mono-oleate, monopalmitate, monostearate and tristearate. Emulsifiers.
E440:	Pectins and pectates. Extracted from fruit skins, these are generally used as gelling agents.
E450:	Sodium and potassium diphosphates, triphosphates and polyphosphates.
E460:	Microcrystalline cellulose.
E461:	Methylcellulose.
E463:	Hydroxypropylcellulose.
E464:	Hydroxypropylmethylcellulose. Gelling agent used especially in fried food.
E466:	Carboxymethylcellulose.
E470:	Salts of fatty acids. Emulsifiers and stabilisers in, for example, packet sauces.
E471:	Mono- and di-glycerides of fatty acids. Very common in, for example, reconstituted dehydrated potato flakes.
E472:	Acetic, lactic, citric and tartaric acid esters of mono- and di-glycerides of fatty acids, etc.
E473:	Sucrose esters of fatty acids.

E474:	Sucroglycerides.
E475:	Non-polymerised polyglycerol esters of fatty acids, permitted in limited doses in, for example, so-called 'fine pastry'.
E476:	Polyglycerol polyricinoleate.
E477:	Propane-diol (or propylene glycol) esters of fatty acids.
E479b:	Soya oil, oxidised and reacted with glycerides of fatty acids.
E481 & 482:	Sodium and calcium stearoyl-2-lactylates. Emulsifiers.
E483:	Stearyl tartrate.
E491-495:	Sorbitan mono- and tristearate, monolaurate, mono-oleate and mono-palmitate.

Miscellaneous

E516:	Calcium sulphate.
E574:	Gluconic acid. Often used in manufactured sauces.
E630:	Isonic acid. Used as a 'shining' agent on cakes and cold meats, but also in fizzy drinks, etc.
E950:	Potassium. Powerful artificial sweetener.
E951:	Aspartame. Powerful artificial sweetener.
E954:	Saccharine. Powerful artificial sweetener.

And remember, there are many more!

SOME RECOMMENDATIONS

In other words, a few publications which bear little or no relation to cookery books.

BRAITBERG Jean-Moïse. *Le Scandale des vins frelatés*. Rocher.

CÉPRÉ Marie-Paule and LEDERMAN Danièle. *L'Année de la lune rousse*. Michel Lafon.

COFFE Jean-Pierre. *Au secours le goût*. Le Pré aux Clercs.

LEDERER Jean. *Encyclopédie moderne de l'hygiène alimentaire*. Maloine.

MARCEL Jean-Claude. *La Sale Bouffe*. Ed. Bernard Barrault.

DE MEURVILLE Élisabeth. *Question de goûts, les bons produits et les autres*. L'Archipel. (A reporter for France-Inter, Mme de Meurville is also Chief Editor of the review *Les Gourmets associés* and author of numerous books on food.)

MOHTADJI-LAMBALLAIS Corinne. *Les Aliments*. Maloine.

NUGON-BAUDON Lionelle. *Toxic Bouffe*. Lattès-Marabout.

DE RONSAY Joël. *La Malbouffe*. Olivier Orban.

VARIOUS AUTHORS. *Nourritures*. Autrement.

Reviews, Magazines and Miscellaneous

Revue de l'Industrie agro-alimentaire.

La France agricole.

Agriculture Magazine.

Cultivar.

Le Canard enchaîné (especially *Dossiers du Canard* No. 6 entitled '*Les dessous de la table*').

60 Millions de Consommateurs.

Que Choisir (in particular the articles by Fabienne Maleysson and Serge Michels).

Not forgetting the *Revue de la Concurrence et de la Consommation* and the annual reports of the *DGCCRF*.

DON'T FORGET THE TIP!

*Without them, our questions would not
have been answered and this book
would not have been possible.
Ladies and gentlemen,
we give you our warmest thanks.*

Emmanuel Baltzer, processed meat producer.

Ter Barseguian of the French Office of Protection against Ionising Radiation.

Annick and Christian Bernard, farmers.

Odette Blanc for her knowledge of poultry farming.

Joseph Bourgeais, farmer.

Claude Chassagnard and Christian Baradel of the Institute of Physics and Chemistry in Paris.

Jean-Claude Evrard of the French Department of Consumption, Competition and Suppression of Fraud *(DGCCRF)*.

Geneviève Fabre of the French Meat Information Centre.

Josiane Fournier of the French Electricity Board.

Antoine Gerbelle, journalist and wine expert.

Georges Gottsegen of the *Marché d'intérêt national de Rungis*.

Jean-Marc Hain.

Bénédictine Hermelin, Secretary of the French Farmers' Union.

Daniel L'hommet, baker.

Jacqueline Marais, aviculturist.

Élisabeth de Meurville for her enlightened advice.

Nathalie Mic Macher of the French General Food Commission.

Ginette, René and Patrick Morin, farmers.

Olivier Navarre of *Française-Maritime*.

Claude Palis, expert mycologist.

Gilles Richard of the French Federation of Processed Meat Producers.

André Rouault of *Cooperl Abattoirs*,

and all those who, for perfectly understandable professional reasons, preferred to remain anonymous,

as well as the *Moulin des Planches* and in particular M. Roland Langlois,

the *Moulins de Cherisy* and in particular M. Yves Lethuillier,

and above all

Mme Lionelle Nugon-Baudon, Doctor of Biochemistry, Toxicologist and Senior Researcher of the French Department of Nutrition, Food and Food Hygiene at the *INRA*.

INDEX

AB logo (agriculture biologique) 166
abattoirs 71, 87-90, 125, 158
abortion 26, 34
acceptable daily intake *see* ADI
acrolein 86
additives 109-22, 134, 140, 168, 182, 198, 201, 207
ADI (acceptable daily intake) 20-6, 51, 55, 62, 108, 110-11, 139-40, 141, 166
    children's 133, 160
    of erthyrosine 114
    and farming 48
    of mercury 34
    of nitrates 53, 54
    of PCP 51
    of sulphites 114
    and 'technological assistance' 100
advertisers 163, 164
aflatoxins 61-3
Africa 36, 59, 94
agricultural groups 22

Algeria 59-60
Alsace, France 29
aluminium 42-4
Alzheimer's disease 43-44, 75, 140
Amazon River 33
American Soybean Association 96
amino acids 86, 131
amylose 118
anabolic steroids *see* steroids, anabolic
anaemia 38
anaphylactic shock 121
'animal by-products' 90-92
animal feed 61, 62-3
animal rights groups 77, 159
animals
    'animal tracing' 78-9, 80
    and hormones 63-5, 71-2, 73, 77, 83, 87
    and steroids 65-9, 72, 73, 77, 87, 91
    transgenic 146-7
    transporting 77, 158-9
anti-bacterials 83, 84

anti-parasitics 83-4
antibiotics 83, 85, 192
AOC (*appellation d'origine contrôlée*) 179, 181
apomixis 147-8
apples 52, 160-61, 206
apricots 50-51
Argentina 58
aromas, artificial 116-20
artificial insemination 71
asbestos 64
ascorbic acid 182
asexual reproduction 148
Asia 36
aspartame 133, 134
Aspergillus flavus bacterium 61
Australia 41
Austria 168

B agonists 155, 156
B-47 (disappearance in the Mediterranean) 31
B-52 bombers (mislaid warheads, 1968) 31
baby food 52, 135, 160
Bardot, Brigitte 158
battery farming 71, 86, 87, 101, 145-6, 159, 190
bauxite 43
Bay of Naples 31
Beaufort Sea 31
becquerels (Bq) 26
beer 182
beetroot juice 113
Belgium 23, 39, 155-6
benzopyrene 36
beta-carotene 135
BHA (butylated hydroxyanisole) 111, 132, 135
BHT (butylated hydroxytoluene) 111, 132
biotechnology 141, 147, 191
blood poisoning 38
Bobigny 193
Bohunice nuclear power station, Slovakia 30
botulism 89, 131
Bourre, Jean-Marie 131-2
bovine spongiform encephalopathy *see* BSE

branding 164, 167
Brazil 31, 57
Brittany 54, 79, 87, 91, 199
bromelin 182
brucellosis 89
BSE (bovine spongiform encephalopathy) 73-81, 83, 91
butanol 118
butter 114
butylated hydroxyanisole (BHA) 111, 131-2, 135
butylated hydroxytoluene (BHT) 111, 131-2

cadmium 41-2, 44
Cahors 68
calciferol 119
calves 63, 68-9, 71, 76, 77, 144, 159
campylobacter 89, 157, 165
Canada 33, 41
cancer 132, 140, 204
and aluminium 43
and ant solution 50
and BHA 111
and BHT 111
and cereal growers 56
and the Chernobyl disaster 29
deaths in France 20
deaths in the UK 20
and food 20
and hormone substances 63
of the liver 62
and nitrates 53
and nitrosamines 131
and PAHs 36
and PCP 51, 160
and platinum 44
and radiation treatment 85
and Seveso 35
testicular 204
Canton 15-16, 17
Cape Cod 30
captan 50
car batteries 41
Carnegie Institute, Washington 146
carob flour 126, 128-9, 133
carotenoic acids 113
carrageenan (seaweed) 104-5, 124, 160

INDEX

Catalogue officiel de l'agriculture 93
catalytic converters 40
cattle breeders
   and 'animal tracing' 80
   Belgian incident 66
   BSE scandal and 78
   calf export 77
   and health certificates 79
   and hormones 64, 65
   and steroids 66
cellulose 137
Cépré, Marie-Paule 53
cereals 47, 49-50, 55-6, 58, 61, 62, 93-4, 96, 97, 141
cesium 134 29
chain reactions 32
Chambery 68
chaptalization 176, 178
Charley-Adele, Jean-Max 156n
cheese 134, 136, 144
Chernobyl disaster 22, 25-9
cherries 50, 114
Chevillot company 80
Chicago 58, 59, 60
chickens 34, 62, 86-7, 103, 125, 144-5, 157, 159, 164-5
children 39-40, 208
   China
   dog meat 16, 17
   gums experiment 105-7, 132
   meat consumption 19
   and pesticides 57
   pigs in 17-18
   and soya 95, 96
   and tapioca 120
   and Wheat Associates 94
   wheat production/imports 59
chinomethionat 50
Chisso factory 33
chloranil 57
chlorophylls 113
chymosin 144
cigarettes 36
cirrhosis 42, 62
citric acid 115-16, 135
citrus red 113, 114
'claw hands' 38
clenbuterol 65, 156

'cloned' vines 174
Clostridium botulinum 89
Clostridium perfringens 89
clouds, and mercury 33, 34
cochineal 113
Codex Alimentarius 71
Colombia 48, 93
colorimeters 115
colourings 111-15, 116, 124, 126, 127, 130, 137, 183, 196
Common Agricultural Policy 60
Condé-Petit, Béatrice 118
Conference on North Sea Protection 175
congenital abnormalities (children of fruit growers) 56
consumer associations 78, 209
corn-flakes 94, 97, 99, 123
Corsica 27
Court of Coutances 155
crab-sticks 103
Creutzfeldt-Jakob disease (CJD) 75, 81, 140
croissants 118, 135
cruise missile (in the Beaufort Sea) 31
Cuba 59
curcumin 114
customs officers 158-9
cyclodextrins 118

daminozide 135
DDT (dichlorodipheyl-trichloroethane) 57
decanal 118
Denmark 23, 26, 39, 41, 157, 168, 204
Department of Consumption, Competition and Suppression of Fraud (DGCCRF) 134, 150, 156, 159, 160, 167, 170, 171, 180, 183, 187, 188n, 189, 193
Departmental Office of Public Sanitation 153
diacetyltartaric acid 135
dichlorodiphenyl-trichloroethane (DDT) 57
dimethylhydroxyfuranone 119-20
Disneyland, near Paris 92
diuron 175

dog meat 16, 17
drinking water, and nitrates 54
Drôme area, France 28

E102 *see* tartrazine
E127 *see* erythrosine
E220 *see* sulphur dioxide
E300 135
E320 111
E321 111
E322 *see* lecithin
E330 *see* citric acid
E338 111
E401 135
E466 135
E472 135
E. coli (Escherichia coli) 89, 90
E Numbers 110-11, 215-24
Eastern Europe 67, 192
eating out 195-201, 207
eggs 101-3, 189-90
embryos, and zinc 44
Erasmus, Desiderius 151-2
erythrosine 113, 114
ethanol 118
ethephon 50
European Agricultural Summit (February, 1995) 159
European Commission 20, 29, 35, 65, 77-8, 149, 168, 169, 175, 192
European Community (EC) 58-9, 65, 69, 75, 81, 109-10, 182, 204
European Parliament 168
European Union (EU) 59, 78
evaporation 34
Evrard, Jean-Claude 27-8
exhaust fumes
  effects on children 39-40
  and PAHs 36
extrusion-cooking 123-37

F-14 missile (in the sea off Scotland) 31
F-102 (in Haiphong Bay) 31
faecal bacteria 89
Farmers' Association 78
Farmers' Union 69-70
farming 45-60

fast food 196
fatty acids 135
female hormones 205
fenchone 118
fertilisers 47, 52, 53-4, 56, 60, 87, 91, 130, 140, 141, 166, 177
Finland 157, 168
fish
  in China 17
  and critical annual fishing limit 34
  and mercury 32, 33
  and nonylphenol 205
  in restaurants 199
fizzy drinks 183
Fla AFR73 137
flavour enhancers 120-21
flavourings 124, 126, 127, 130, 135, 183
flumequine 84
fluorine 44
food allergies 121-2
food chain 20, 36, 60, 62
food poisoning 90, 189, 201
Food Sciences Centre, Zurich 118
France
  abattoirs in 90
  and aflatoxin 62
  agricultural foodstuffs as premier industry 13
  animal identity tags 80
  and cadmium 42
  cancer deaths in 20
  and Chernobyl 26-9
  and the cochineal 113
  constituency boundaries 23
  Department of Food 24
  and eggs 102
  farmers 55, 57, 60
  and food and health documents 23
  fruit growers 56
  and grain 93
  and hormone-treated meat 64
  and hormones 72
  investigations in 153
  and knackers 91
  and lead pellets 41
  meat consumption 19
  mercury content of fish 33
  Ministry of Agriculture 57

INDEX

and nitrates 53, 54
and PCP 51, 52
radiation levels in 26
radioactive waste containers lying off the French coast 32
and the 'Seveso directive' 35
and soya 95, 97
and steroids 67
and sugar 103
'traditional national products' 168
and water 54
wheat production 59
wine growers 52
Francescella tularensis 89
frankfurters 130
Fraud Squad (France) 154, 180, 186, 188, 193
'free-range' eggs 159
French Atomic Energy Commission 26
French dressing 115
'freshness tags' 189
frogs' legs 85
fruit
  and aroma additives 117, 119
  in baby food 134-5
  and fungicides 47
  hybridisation 143
  and insecticides 50
  and liquid nitrogen 50
  and mutagenesis 141
  washing 55
fruit boxes 51-2
fruit farming 50-51
fruit growers 56
fungicides 47-8, 50, 160

gamma dodecalactone 119
gamma rays 161
gastroenteritis 140, 157
GATT (General Agreement on Tariffs and Trade) 72, 192
gelling agents 128, 129
genetic manipulation 22, 140-42
genetic transformation 146-7
genome project 210
geranyl acetate 117
germanium 44
Germany 23, 26, 51, 157, 168, 182

glutamate 120, 121, 130
glycerol 178
goats 147
Golfe du Lion 32
grain 93-5
grays (Gy) 26
Great Britain
  and Chernobyl disaster 28
  EU blockades over BSE 78
  food and health documents 23
  meat consumption 19
  slaughter of cattle (1989) 76
  slaughter of cattle (1996) 78
  and transporting of animals 77, 158
  veal offal ban (1995) 76
Great Lakes 33
Greece 168
'green paper' on food laws (EC) 168-9
greenhouse effect 34
growth hormones 63
guar 126, 133
Gueret 70
gums 105-8, 124, 125, 126, 128, 132

Haiphong Bay 31
ham 127-9, 133, 134, 152
Hammourabi Code 151
Hartning, Professor Nolan 72-3
Hasegawa Research Centre, Japan 119
heavy metals 37, 87
herbicides 174, 175
herbs, and Chernobyl disaster 28
Hermann (transgenic bull) 147
hexanal 118
hexanal 2.4 decadienal 119
Hites, Ronald 57
Hohenheim, Theophrastus Bombastus von 21
Holland 23, 28, 41, 72, 155-6
Hong Kong 15
hormones 63-5, 71-2, 73, 77, 83, 87, 144, 155, 191, 192, 207
hunters 40-41
hybridisation 143
hyperactivity, and lead pollution 40
hypodermasis 79

Ignalina nuclear power station, Lithuania 30
imports 156-8
impotence 56
INAO (*Institut national des appellations d'origine*) 179
INRA (National Institute of Agronomic Research) 41, 89, 146
insecticides 46, 48, 50, 52, 57, 63, 140
INSERM 131
Institute of Physics and Chemistry, Paris 111
Institute of Scientific Research for Development Through Co-operation (ORSTOM) 147
insurance companies, and Seveso 35
intelligence, and lead 39-40
International Safety Organisation (ISO) 201
investigations 153-4
Iran 48
irradiation 85
ISO (International Safety Organisation) 201
Italy 26, 27, 73, 113

Japan 57, 94, 103

Kara Sea 32
kidneys, and cadmium 42
knackers 90, 91
Komsomolets Soviet submarine (sunk off the Norwegian coast) 31
Korea 103
Kozloduy nuclear power station, Bulgaria 30

labelling 24, 121-2, 149-50, 152, 159-60, 168, 170, 178-9, 180-81, 191-2, 193, 207
lactoflavin 114
lactones 119
latex 128
Lauvergeon, Anne 30
law, the 150, 155, 168-9, 171

LD (lethal dosage) 21-2
lead 44
  and children 39-40
  diseases caused by 38-9
  in the home 39
  and hunting 40-41
  and hyperactivity 40
  and intelligence 39-40
  maximum permitted dose in food 38
  and retarded growth 40
  and tap water 183
'lead colic' 38
lead pellets 40-41
lecithin 135, 137, 192
Lenoir, Noëlle 149
lethal dosage *see* LD
life expectancy 203
'light' ('sugar-free') craze 169
limonene 118
lindane 57
liquid nitrogen 46, 50, 53, 54
listeria 154
Listeria monocytogenes 89
liver, and cadmium 42
liver cancer 62
livestock health certificates 79
Los Alfraques campsite, Spain 86
'low-fat' products 169
luteotropic hormones 144

McGill University, Montreal 144
'mad cow disease' *see* BSE
Madagascar 31
Mao Tse-tung 105-6, 132
mayonnaise 124
meat
  Belgian 66-7
  fall in consumption of red 67
  imports from Eastern Europe 67
  mechanically separated 129-30
  overproduction of 67, 69
  premier quality 69
  processed 127-31, 134
  in restaurants 199
  second-rate 69
  'tracing' 78-80
meatballs 124-6
mechanically separated meat (MSM) 129-30

INDEX

menthone 118
mercury pollution 32, 33-4
methionine 86
methyl bromide 50
Mexico 48
Miami 65
  milk 157, 190-92
  and Chernobyl disaster 38
  and lead pollution 40
  production quotas 67
Minamata 32, 33-4
minerals 169
Ministry of Agriculture (France) 164, 166
Ministry of Consumption (France) 164
Mitterrand, François 30
motor damage 38
MSM (mechanically separated meat) 129-30
mutagenesis 141-2

naphthaleneacetic acid 50
naphtol 118
National Cattle Federation (France) 65, 69
National Centre for Veterinary and Alimentary Studies 153
National Federation of Agricultural Enterprise Syndicates (FNSEA) 210
National Food Hygien Commission 20
National Institute of Agronomic Research (INRA) 41, 89, 146
nectarines 143
nervous system
  and cereal growers 56
  and exhaust fumes 39
  and germanium 44
  and glutamate 121
  and mercury 34
Netherlands see Holland
'new products' 97, 99, 125, 136
nitrates 53, 54, 55, 166, 183
nitrites 128, 131, 166, 175
nitrosamines 131
nonylphenol 204-5
Noppen, Karel Van 66
Norway 26, 31, 32, 157

nuclear power stations 30
  Chernobyl see Chernobyl disaster
nuclear reactors (dumped into the Kara Sea) 32
nuclear satellites (fallen into the sea near Brazil and Madagascar) 31-2
nuclear submarines 30, 31
Nugon-Baudon, Lionelle 41

obesity, and meat 63, 72
octanal 118
oestrogen 63, 204
oleic acid 119
omatotropin 72
'organic' products 166-7, 208
osteosclerosis 44
overpopulation 205
oxidisation 188
oxine-copper 50

PAHs (polynuclear aromatic hydrocarbons) 36
Paks nuclear power station, Hungary 30
pancakes 135
papain 182
Paracelsus 21
parathion 48-9, 50
Paris (child deaths from lead poisoning) 39
patulin 160
PCBs (polychlorinated biphenyls) 35-6
PCP (pentachlorophenol) 51, 52, 160
peaches 143
pectin 160
pentachlorophenol see PCP
pesticides 47, 53, 55, 57, 60, 63, 87, 130, 140-43, 166, 204, 208
pet food 91-2
petrol, 'unleaded' 40
PHB (polyhydroxybutyrate) 146
phosalone 50
phosphates 127
phosphoric acid 47, 50

pigs
  in China 17-18
  and fish meal 34
  in France 19, 62
  slaughter of 87-8
  sterilising 147
  transgenic 146
pizzas 135
plastics 111, 146, 204, 210
platinum 44
poisons (lethal dosage (LD)) 21-2
Poitiers 68
politicians 209
polychlorinated biphenyls *see* PCBs
polyhydroxybutyrate (PHB) 146
polynuclear hydrocarbons *see* PAHs
polyphosphates 128
pork 157, 165, 206
Portugal 26, 73, 113
potassium 47, 50
potassium sorbate 135
premature senile dementia *see*
  Alzheimer's disease
preservatives 130, 211
processed meat 127-31, 134
profit motive 21, 23, 71, 72, 77, 132, 205, 208-9
propylene 86
propylene glycol alginate 182
protective clothing, and farming 48, 56
proteins 128, 136
Public Hygiene Council 132

quails 145-6
Que Choisir ('*Which?*') (French Consumers' Association) 63, 64, 67, 132, 157, 165, 189
Quincke's edema 121
quinoline 114

radiation
  effects on the body 26
  levels in France 26
radioactive waste containers 32
radionuclides 87
'Red Label' products 69, 164-5
refrigeration 185-93
'rendering' factory 91

Rennes 54
rennin enzyme 144
Research Centre for the Study and Observation of Living Standards (CREDOC) 169
research and development departments 126-7
resins 128, 129, 130
retarded growth, and lead pollution 40
Revue de l'Industrie agro-alimentaire 119
riboflavin 114
Rocard, Michel 64
Rovno nuclear power station, Ukraine 30
Russia 58

Saint-Blanquat, Georges de 132
salads, pre-packed 85, 105
salmonella 89, 154, 157, 186
saltpetre 127
saturnism 38-9
sausages 152
Scotland 28, 31
seaweed 105, 124, 133
seeds 45-6, 52, 62, 141-2
Seehofer, Horst 81
selenium 44
sell-by date 189, 211
'semi-processed' products 137
'set-aside' policy 58, 59
Seveso, Milan 34-5
'Seveso Directive' 35
sheep 28, 74
sieverts (Sv) 26
simazine 175
Single European Market 157
slaughterhouses *see* abattoirs
slimming products 169-70, 171, 205
'smart-tag' 80
smoking 36
'soil treatment' products 46
Sophia-Antipolis, France 28
South America 36, 65, 93, 113
South Korea 44
Soviet Union, former 59, 65
soya 95-7, 127
soya bean flour 86
soya cakes 62, 95

INDEX

soya oil 96
soya steaks 97, 105, 112, 123, 124
Spain 26, 27, 31, 51, 53, 67, 73, 101, 113, 156
spectrocolorimeters 115
speed limits 40
SSN-589 nuclear submarine (sunk off the Azores) 31
SSN-593 nuclear submarine (sunk east of Cape Cod) 30
Staphylococci 137
starches 136
state controls 151
steel 44
sterility 56
steroids, anabolic 65-9, 72, 73, 77, 87, 91, 155
strawberries 50
sugar 103-4, 128, 134, 176, 178
sugar beet 104
sulphites 114, 121, 175
sulphur 50
sulphur anhydride 133
sulphur dioxide 114
swallow's nest 16-17
Sweden 23, 26, 43, 168
sweeteners, artificial 133, 134, 169
Switzerland 26, 28, 51, 81, 158, 159

Taiwan 40, 44
tapioca 120-21
tartaric acid 178
tartrazine 113, 114
TCDD (tetra-chlorodibenzo-p-dioxin) 34-5
'technological assistance' 100-1, 103, 140, 199
tendonitis 56
tetra-chloridibenzo-p-dioxin (TCDD) 34-5
tetrasul 50
Thailand 62
Thatcher, Margaret, later Baroness 74
thiophanate-methyl 50
thiram 50
'tin' cans 43-4
tine 44
tomatoes 142, 193
'total treatments' 47

trace-elements 169
traffic jams 39
transgenic animals 146-7
tremors 38

UCCS 136
Ukraine, and Chernobyl 26-7
United Kingdom *see* Great Britain
United States of America
   Department of Agriculture 96
   farmers 56-7
   and genetic modification 141-2
   and grain 93-4
   and hormones 72
   and hyperactivity 40
   and lead pellets 41
   mercury in the Great Lakes 33
   and obesity 72
   opinion of European agriculture 72-3
   and PCBs 36
   research in 44, 56
   and soya 95-6
   and sulphites 114
   and wheat production 58, 59-60
Universal Oil Products 111
University of Texas 147
'unleaded' petrol 40
urea 50

vacuum cooked meals 85
vanadium 44
Vaucluse 193
VDQS (*vin de qualité supérieure*) 181
veal 63, 64, 76, 207
vegetable oil/fat 135
vegetables
   cultivation for mass food 206
   dehydrated 85
   and fungicides 47
   and mutagenesis 141
   washing 55
Vidal 84
vitamins 169, 170
Vosges, France 29

THE RUBBISH ON OUR PLATES

water 54, 166, 183
weed-killers 46-7, 50, 175
Wheat Associates 93, 94
wheat production 58-9
white corpuscles 26
white wine 121
wine/wine growers 38, 52, 173-82
World Grain 93

World Health Organisation 175

xanthan gum 133

yoghurt 159-60, 171, 192

zinc 44